The Thin Line Between Heaven and Hell

```
F HAR
Harris, David G.
The thin line between
 heaven and hell /
GUELPH PUBLIC LIBRARY
```

Removed from Inventory
Guelph Public Library

by David G. Harris

Copyright 2000 by David G. Harris
All Rights Reserved

Harris, David G.
ISBN: 0-7443-0273-0
The Thin Line Between Heaven and Hell/
David G. Harris – 1st ed.

No part of this publication may be reproduced, except in the case of quotation for articles, reviews, or stores in any retrieval system, or transmitted in any form or by any means, electronic, mechanical, photocopying, recording or otherwise, without written permission from the publisher. For information regarding permission, contact:

SynergEbooks
1235 Flat Shoals Rd
King, NC 27021
1-888-812-2533
"http://www.SynergEbooks.com"

Contact the author via email at
"mailto:dharris518@hotmail.com"
This book is also available in digital format at SynergEbooks

Cover art by David G. Harris
Graphic Design by ETS Designs

The Thin Line Between
Heaven and Hell

Guelph Public Library

Author Disclaimer:

The story you are about to read is based on actual calls from many different fire departments. The names have been changed to protect the innocent. As far as this author knows, there is no Mannington Fire Department. Any similarity to another person, firefighter, or fire department in the United States or abroad is purely coincidental.

~David G. Harris
   Winston-Salem, North Carolina

DEDICATIONS:

To my beautiful wife, Theresa.
Without her help and support, this book would not have happened.

To Joseph Aaron and Ashley Nicole Revalee
For being the two greatest kids in the world.

To my family
Without their help and support, I would never have joined the fire department.

To the Tetter and Worley Families
Without them, I would never have found Theresa.

To the High Point Fire Department
Without them, I would not have been able to write this book.

To all of the men and women of the fire service that I have met over the years. Without them, this book would not have been possible.

"I have no ambition in this world but one,
and that is to be a fireman.
The position may, in the eyes of some,
appear to be a lowly one;
but we who know the work which a fireman has to do
believe his is a noble calling.
Our proudest moment is to save . . .lives.
Under the impulse of such thoughts,
the nobility of the occupation thrills us
and stimulates us to deeds of daring,
even of supreme sacrifice."

~Edward F. Crocker
Chief of Department
Fire Department New York
1899 - 1911

Introduction

The history of the fire department dates back as far as 24 BC with the Roman emperor Augustus instituting a corps of fire-fighting watchmen and a department consisting of men using buckets and axes. The next known fire brigade came in 1666 following the Great Fire of London. Insurance companies formed fire brigades and the government was not involved until 1865 when London's Metropolitan Fire Brigade was formed.

In the United States, fire wardens were established in New Amsterdam - New York City - and the beginnings of the first public fire department was established in North America. Starting with fire wardens and bucket brigades, fire departments and fire fighting has developed to the present day version by adding tools, pumps, horses, mechanical apparatus and knowledge.

When fire departments were first developed, knowledge of fire behavior was very limited. Today, the knowledge of what causes a fire, what will burn and what will not burn, how a fire will react to the design and construction of buildings has grown. When fire departments were first established, most buildings were made of wood and there were limited exit routes from buildings. Today, with present day fire codes, buildings are safer to live and work in and fire exits are clearly marked to aid in saving lives.

On October 1871, two of the worst fires in the United States happened in Chicago, Illinois, and Peshtigo, Wisconsin. The Great Chicago Fire took the lives of 300 people, destroyed and area five miles long and one mile wide and destroyed 17,000 homes. On the same night, 250 miles away in Peshtigo, population 1,700, fire burned the entire city to the ground. Not only was every home and building in Peshtigo destroyed, more than 1200 people were killed because of the fire. The ones who survived were the ones who jumped into the waters of the Peshtigo River.

Today, with the advancement of knowledge and apparatus, fires of the same magnitude are seldom heard of. Better equipment, better knowledge and the dedication of more than 1.2 million firefighters in the United States allows us to live in safer homes, work in safer environments, and when a fire does occur, we have a better chance of survival.

On a hot day in July 1900, the city I work for - High Point, North Carolina - suffered one of the worst fires in its history. The fire began in an old steam laundry and destroyed practically an entire block in the heart of town. In 1901, the High Point Fire Department came into being out of necessity. The first pieces of apparatus purchased were a hose reel, 500 feet of hose, axes and nozzles.

Today, the High Point Fire Department has eleven stations with more on the way. The fire department has grown from sixty volunteer firefighters to more than 180 paid firefighters. It has grown from a single hose wagon to eleven front-line pumpers, three of which can double as ladder trucks, two front-line ladder trucks, and two squad trucks.

I joined the High Point Fire Department on April 4, 1994. In more than seven years with the fire department, I have experienced one of the most rewarding careers than I thought was possible. I have seen the good and bad, life and death. One of the most exciting times I can remember was the night that I helped bring a life into the world. One of the worst times I can remember, and unfortunately cannot forget, was the day that an eight-month-old baby died in my arms. I have helped people salvage what they could and move out of their burned homes two days before Christmas. And I have helped people who have lost loved ones in their family. All in all, it's been a rewarding career. I may cry for a person or I may be pissed off at someone that has done something stupid and hurt or killed others. But I can think of no other job that I would rather be doing.

What you are about to read is not about one individual firefighter. It is a conglomeration of stories I have heard or experienced myself. These stories have been rolled together into one firefighter. I do not specify which incidents I experienced first hand or heard about through the grapevine.

One reason I wrote this book is to tell the great stories I have heard since I started my career in the fire service and to tell about the things I have experienced first hand. Another reason I wrote this book is to let people know that the fire department is a lot different than what they see on television or in movies.

A lot of people think that all we do is sit around the station playing checkers and cards, just sitting on our butts until a fire call comes in. Nothing could be further from the truth. We train; we ride emergency medical calls, hazardous materials calls, car wrecks, investigation calls, and the like.

Some people look at firefighters as lazy because most of us we only work a third of the year. They don't know the kind of work we have to do and they don't realize the fact that we have to work a second job just to make ends meet. They don't think of the amount of physical training we do and the number of classes that we have to take both on and off duty. If they knew exactly what we did every day, they would realize that this job is a lot different than what they see on television.

Some people don't understand. I tell some about a bad call that I had the night before and all they can say is, "That's the job that you chose to do." That really doesn't help much. I know this is the career I chose, but sometimes you have to talk about your problems with someone who will at least listen to you. That's one of the things I love about my wife. If I have a bad call, I know that I can call my wife, Theresa, and talk things out with her. She is the anchor in my life. She helps me in more ways than I can possibly say here. But the one thing that keeps me going is to know that I can turn to her for any reason. She is always there for me. And for that, I love her dearly.

Hopefully this book will help people understand that the firefighters in their community are there to help them. Contrary to what most city councils would have you believe, we are not in this job for fortune and glory. We are here to serve the public. We will never get rich doing this job. Those who do get rich usually get investigated. We are here because of a love for the job and to help people. We are not adrenaline junkies. Adrenaline junkies can get you hurt or killed. We don't care if we ever get our names in the newspaper. We don't want to be immortalized by the public. We just want to help. Plain and simple.

A lot of the time, we have to fight for what we need. Equipment, more stations, more personnel, and more money. Almost all of the fights are with the city council. They would have the public believe that firefighters, EMS and police personnel are overpaid and do not need more money. They would have the public believe that we are expendable and if we were to leave, public services would not be affected. Of course, when the call affects one of the members of the city council, we are irreplaceable and deserve more credit and money than we are paid. That is, they believe until the next election.

We could make more money working in another field, but there is an unexplainable love for this job that draws all of us here. Maybe one day someone will be able to explain it better than I can. But until then, just know that we are here for you. Call us when you need us and we will be there.

Enjoy,
David G. Harris

"There is a thin line between heaven and hell.
I cross that line every day.
I am a firefighter."

David G. Harris
High Point Fire Department

Chapter One

It was hot that day. When the alarm came in at 8:15 that morning the temperature was already over 90. I had just come back from a nice vacation with the wife and kids and I did not really want to play firefighter in that heat. Besides, I hadn't even had my first cup of coffee.

We could see the smoke as soon as we left the station. Thick black columns of smoke rising up into the sky and leveling off at the top like a mushroom. The address didn't register at first but when we saw the first explosion we knew it was the tank farm. The tank farm contained millions of gallons of gasoline and we knew if it caught fire that it would be hell to put out.

Bullet looked over at me and said, "Welcome back, Cap!"

All I could come back with was, "Kiss my ass!"

We lay in from the hydrant at the gate. All 800 feet of five-inch hose. We hooked the top mount deluge gun with our foam system and started laying a blanket of foam on the east side of the tank. The succeeding engines started laying foam from the north, south and west side. Ladder 8 and ladder 16 came in and set up their aerials and started cooling the other tanks.

That's when we were notified about two men trapped in a building on the east side of the tank. They were pinned in the building from one of the explosions after a utility pole fell and blocked the door. The heat was still building and we knew they would cook if we didn't get them out of there.

I took two firefighters, our chain saw and a hose line with me and started making a path to the building. We were able to make it to the building but found that the pole was not the only problem. A transformer had jammed into the door. We would still be unable to get the door open. We made our way around to the backside of the building and had to put out a few spot fires along the way. We banged on the wall and shouted for the two men inside. "Hey! Anybody in there?"

"Yeah! We're in here but you gotta get us out. Bill's not looking too good and there's a lot of smoke!"

I told them, "Move towards the door! Cover up as good as you can!"

I then had the two firefighters with me keep an eye on the fire as I started the chain saw. I cut a section of the sheathing on the wall, being careful not to cut through the studs. I then kicked the side of the cut and the wall opened up. I went in and grabbed both men. As I pulled them away from the building, the two firefighters covered us with water to keep the two men cool. We made it back to the truck and the men thanked us.

Several hours later, the fire was out and we were taking our lines up. It was a hell of a fight but we had won the battle and killed the beast once again. Nothing special. Just a lot of sweat and hard work. The whole time we were rolling up the hose, all I could think about was the day before. There I was. Lying on the beach with a beer in one hand and rubbing suntan lotion on my wife, Teri, with the other. We were lying there watching our kids, Joey and Nikki play in the water. Then having to come home and visit hell once again. But, then again, there's a thin line between heaven and hell. I cross it every day. I am a firefighter.

Don't get me wrong, I love my job. I wouldn't trade it for anything in the world. I've been doing this for twenty years and I still love it as much as I did from day one. It's just that this particular day I didn't feel like doing it. I just had to get back in the groove.

We got back to the station and cleaned up. We had a new batch of rookies fresh out of training that I had not met yet. We had two assigned to our station and it was time to meet them.

The rookies had come on the shift before, while I was on vacation. The guys had told them stories about me. They told them that I like to run a tight ship. Bullet told them I was ex-military and I liked for my rookies to snap to attention every time I passed them. Being new and nervous, these two fell for it, hook, line and sinker. When I walked in, both of them jumped out of their seats, standing at attention. I looked at them, not exactly sure what the hell was going on. I walked up to the first one and asked his name.

"Johnson, sir. Rick Johnson!" he said.

"Where you from, Johnson?"

"Right here in Mannington, sir!"

I looked around the room at the other guys. I knew something was up. They could barely hold back the laughter. I stepped over to our other rookie. "And what's your name?"

He snapped to attention, "Williams, sir. Steve Williams!"

"And where are you from, Williams?"

"Huntington, West Virginia, sir!"

At this point, I'm thinking these are two of the biggest suckers I've seen in a long time. "Huntington, West Virginia, huh? Are you married, Williams?"

"No, sir!"

"Got any brothers or sisters?"

"No, sir. I'm an only child."

"That's too bad. I guess that means you're going to die a single man, doesn't it?"

The room fell apart at that point. Everybody was rolling with laughter except for the rookies. They didn't know whether to take me seriously or not. You could tell that they were not sure whether this was a setup or not. That's when Bullet came clean about me. About how I am one of the most laid back captains on the department. I have to hand it to Bullet. He really had those guys going. They didn't know what they were in for.

Bullet's a good guy. He has a short temper but he's still one of the best drivers I have ever seen. He came on this department in the same rookie class that I was in. Richard Arthur Stack was only 20 years old when we came out of training. He had an old clunker of a truck that would only run half the time. On his second shift, he couldn't get the truck started and was late getting to work. He was docked some pay, had several letters of reprimands put in his files, and he was chewed out by two battalion chiefs, a deputy chief, and the chief of the department. He was having one of the worst days of his life. All he could do all day was mutter to himself about that piece of shit truck of his. The next morning, I gave him a ride home. When we pulled into his driveway he told me to wait there. Bullet walked into his house and came out a minute later with a gun. I didn't know what I was in for. Bullet walked over to his truck and proceeded to fire six shots into the motor. He walked over to my Jeep, smoking gun in hand, and asked, "Give me a ride tomorrow?"

"Not a problem," I said. I could only sit there in disbelief as I watched Bullet walk back into his house and close the door.

The next day I told the other guys what had happened. That night there was a single bullet lying in the pillow of Bullet's bed. Ever since then he has been known as "Bullet."

Chapter Two

To do this job, you need to have three qualities. First, you have to be willing to put your life on the line for others. To do whatever it takes to save a life, even if that means you may lose yours in the process. Second, you have to be willing to overcome your fears. To know that at any second the fire could come out and bite you, that a wall could collapse on you, that a floor could cave in with you. You have to know your fears and you have to be willing to overcome them. The third, and perhaps the most important quality of a firefighter. . . . you have to be a little nuts.

It doesn't take a genius to be a firefighter. Lord knows people wouldn't think we were very bright since we run into burning buildings when everybody else is running out. No you don't have to be the smartest person in the world, but there is one thing that separates firefighters, police officers and EMS personnel from the rest of the world. Heart.

Heart gives us the courage to do our job without thinking about what may happen to us. Oh, we know in the back of our minds that bad things could happen. We know that at any second we could lose our lives. But we have to push that to the back of our minds so that we can think of what is happening at that very moment. That moment when we are facing the fire, or as we like to call it, the beast. Heart also gives us the ability to love a little more, to care a little more, to play a little harder, and to live a little better. Heart lets us see the humor in life in a way that no one else can see. That humor may, in the eyes of others, seem a little harsh, but it's a way to let some of the stress of the job.

People may look at firefighters as lazy. I have heard people say that all we do when we are on duty is sit around playing cards or playing checkers and just wait around for a call to come in. I wish to hell that I could find a department like that. People don't realize that most fire departments in the United States are now running emergency medical calls and that we have to train for those calls. They don't realize that we have to take other classes for fire ground training, hazardous materials, and all other types of emergency calls. If the truth be known, firefighters spend more than a third of their careers training for what may happen whether it ever happens or not.

People think that we only work a third of the year. There are three shifts in most cities and they ordinarily work twenty-four hours on the job and have forty-eight hours off. That sounds good to most people. But the fact is that they don't realize that firefighters are usually not paid very well. They don't know that most firefighters work a second job on their days off just to make ends meet. When the public sees our schedule and they think about how great it would be to only work a third of the year, they don't think about the fact that we work on holidays, birthdays, and special events. They don't think about the fact that we are here working when they are at home celebrating Christmas with their families. They don't think about the fact that when it's their son or daughter's birthday, they can be at home celebrating it with them. We have to be at work and celebrate a day later. The fire department never closes. There is someone on duty twenty-four hours a day, seven days a week. We may only work a third of the days in a year, but we also work a third of the holidays as well.

People don't realize the stress that this job brings to your life. Any time I talk about the stress in the fire department, they always say, "Oh, all jobs are stressful. What makes your job more stressful than mine?" The job is physically stressful. You have to pull heavy hose through buildings. You have to be able to bend that hose, which is stiff as a board, around corners. You have to be able to carry that hose upstairs, downstairs, across long distances. They physical stress of this job is tremendous.

The mental stress is twice as bad. I once had an insurance agent tell me that my job had less mental stress than her job. I couldn't believe she would say something like that. I had to explain to her that if she screws up on her job, somebody's insurance policy gets canceled. If I screw up on my job, somebody dies. There's a big difference there. An insurance policy getting canceled is nowhere near as bad as someone's life getting canceled. Of course, after explaining this, she informed me that this was the career that I chose. Well, no shit Sherlock. I know that. Then I reminded her that if I wasn't there when the public needed me, she wouldn't be able to pay all of those big insurance claims and wouldn't have a job.

People wonder what would make anyone want to be firefighter. Dealing with people who have lost their homes or loved ones. Dealing with sick people. And, more and more these days, dealing with people who would love nothing more than to kill you. Most firefighters can't answer that question in a single thought. Most just say that they want to help people. And, while that is true, there is something more to their reasons for wanting to be a firefighter. It's hard for most of us to pinpoint, but if there is one thing that means more to a firefighter than anything else, it would be brotherhood. There is a brotherhood among firefighters that is like no other. Each of us knows what the other has been through even if we were not there to go through it with them. The death, the destruction, the loss. We've seen it, too. Even the probies, or probational firefighters, feel the brotherhood. Every firefighter at one time was a probie. We know what they are going through. Of course, that doesn't stop us from giving probies hell, but we still know. We had to catch hell from the veteran firefighters, too.

Being a firefighter is like having an extended family. All other firefighters are our brothers and sisters. Even when we travel we can go to another city's fire stations and feel just as at home as we do in our own stations. We can talk to firefighters from other cities and know what they have to go through even though we have never met them. In short, even though fire departments have different administrations, different leaders, they are still the same throughout the United States. We have to deal with loss. In some cases the calls can be humorous, but more often than not they are about loss.

In the years that I have been with the fire department, I have seen plenty of fire. I have seen the beast and looked him in the eye. I have battled with him. He has bitten me. I have scars on my body from my years with the fire department. But I wouldn't trade those scars for all of the money in the world. I would have preferred to not be injured, but that's part of the job. The scars are my way of remembering that the beast can bite you. The beast has only nipped me, but that instilled in me a greater respect for the beast and what it can actually do. That respect keeps me from getting careless.

The fire department has always been good to me. I don't make a lot of money and I don't lead a glamorous lifestyle but I get by. Of course, I have to work another job on my days off but I would not trade my job for any other job out there.

When I started with the department, I had no idea what I was in for. I just knew that a job in the fire service is what I really wanted to do. I guess it goes back to my teen years watching shows like Emergency. I used to get a thrill every time that show came on. I still love watching the reruns. It showed me what I wanted to do with my life. I graduated college and worked in a different career field but I was still drawn to fire fighting like nothing else I had ever known. Then one day, about five years after graduating, I saw an ad in the classifieds telling that the fire department was taking applications for new firefighters. I applied and was finally hired two years to the day after applying for the job. I had all but given up hope when one day, out of the blue, the chief of the department called me and set up an interview. Two days later, I had the job I had always dreamed of.

    I was nervous as hell the first day. We met at headquarters station and had to get our turnout gear and uniforms. We had to fill out all the paper work and meet all the chiefs. Then we went out and started our physical training. I wasn't in the best shape of my life and the physical training was hard for me. Fortunately, it was April and still cool enough so I didn't look as close to death as I actually felt. But eventually I got to where I could run with the rest of the rookies and I started feeling better about that.

    The first day I actually went into a fire was nerve-racking as hell. I knew then that this would either make me or break me. I knew it was going to be hot. I knew that I would be going into a dangerous situation. But I also knew that I would be going into that fire with experienced veterans. I knew that they would never let anything happen to me. The trainers took us into a smoke building and started a fire in straw with the corner of the room. They wanted us to see how a fire built up from a spark and would climb the walls of the room looking for more oxygen. They showed us how the fire would roll across the ceiling and start banking down the wall on the opposite side of the room. The room started getting extremely hot at that point, but as I watched the fire rolling across the ceiling, I thought to myself that there could be nothing more beautiful than that. Dangerous, but beautiful. From that point on I was hooked.

Fire searches for food in what it burns. It generates heat. It looks for oxygen. When it can find no oxygen in the area, it moves further, always searching. It lives. It breathes. It eats anything in its path. It grows. Fire is a living, breathing animal. It is a beautiful animal. Tamed, it provides heat for homes and cooking. Untamed, it is deadly. Knowing this, is it any wonder that firefighters call it the beast?

I've fought the beast in all kinds of weather. In the summer, you learn to hate the beast. The heat is already high. Add to that the turnout gear and the heat from the fire and it is intense. In the fall and spring, fighting fire is not that bad. Sure, it's hot in the fire, but when you come outside it's nice. In the winter, you come to welcome the heat. It is a welcome change from the frigid temperatures outside. It's almost like mixed emotions when the fire is out. You are happy that you have stopped the fire but you also know that you have lost the only source of heat on the scene. Any exposed skin is numb. Your hands and your face are numb. Your feet burn from being in the cold. I have even come in from a fire in the winter and had to stand in front of the oven with the door open to thaw my bunker pants out enough to take them off.

But with all its faults, the job is still the best in the world. I've seen death, but I have also seen life. There is nothing like the look on a mother's face when they see their child for the first time after delivery. To see the look on a parent's face when you pull their child out of a fire and deliver them to their parents' arms. That's what makes this job worth it.

There are also the calls that make you laugh. Sometimes you just can't help it. Like the time we ran a call where a man was stuck in the bathroom. When we got the call, we thought that it was nothing more than the door being stuck or the doorknob falling off. When we arrived, the man's wife told us what had happened. This man would come home from work at three thirty every afternoon. When he came home, he would always take the sports section from the paper into the bathroom and stay there until he read the entire section. Apparently, this particular day's sports section was longer than usual. The gentleman had sat in his bathroom until his legs fell asleep. When he went to get up, he stumbled backwards and fell between the toilet and the bathtub. He was a rather large man, weighing in around 260 pounds. When he fell, he ended up with his butt stuck between the toilet and the tub.

When we arrived, I looked in on him as his wife was talking to my captain. When I saw him, I was doing a pretty good job at holding the laughter in. The bathroom was at the end of the hall and from the mirror you could see all the way to the other end of the hall. As I sat the toolbox down on the counter in the bathroom I looked in the mirror and could see the man's wife with my captain and driver just rolling with laughter. That's when I lost it and started laughing, myself. The gentleman became very upset and started swinging at me. But one threat of leaving him there in his predicament quickly calmed him down.

Upon further inspection, and from the man's story, I found out that the man was seriously stuck between the toilet and the tub. His family had tried everything they could think of to get him out. They had poured cooking oil, shampoo, dish soap, water and baby oil on him trying to get him to slip out. His wife and two sons tried pulling him out to no avail. Finally they called the fire department. They did, however, allow the man to keep some of his dignity. They could not get his pants back up but they were able to take two bandannas and tied them around the man so that he would not be totally exposed. Thankfully. We ended up having to drain the toilet and remove it from the floor to get the man out. And, of course, being the rookie, I was the one who was nominated to crawl under him to get the bolt out of his side of the toilet. I guess it is fortunate that the man had finished his business before he fell. We removed the toilet and, even though the man had been in such an embarrassing spot, he was extremely good-natured about the whole incident. Messy as hell and extremely embarrassed, but nonetheless, good-natured.

Chapter Three

One of the things that make the job worth it is the time spent in the firehouse. These can be some of the best times in a firefighter's career. That is one of the things the retirees say they miss most about the job after retirement. Time spent with the rest of the firefighters.

There are a lot of jokes played on firefighters in the stations. Other companies coming in and short sheeting beds while the company is out. Rigging beds to fall when someone tries to get in them. We have put cellophane on the toilets at other stations. We have rigged books on nightstands to fishing line. In the night, you pull the fishing line so that the book will be hovering over the person's face when he wakes up. This has a tendency to freak a person out at night. Another popular trick is to tie fishing line to a person's blanket. You pull it and the covers come down. Of course, when this trick has been found out in the past, it has been turned around.

We had a rookie try this trick one night on a long-time veteran of the department. He worked for an hour setting up the line and hiding his work. He then went to the day room to watch television. While he was out, the veteran came in, spotted the fishing line right away and added more line leading back to the rookie's bed. That night, the rookie came to bed and started pulling the line. As he would pull the line, everyone heard him giggling and in the dim light could see him pulling his own blanket back up. This went on for a few minutes. Finally, after the rookie had to pull his blanket up for the tenth or eleventh time it dawned on him that he was had. He is now the veteran but no one has let him forget about the night his trick backfired.

Another popular trick is to tie party poppers to beds. Party poppers are small firecrackers that have a string on each end. The idea is to tie one end of the popper to the fitted sheet and the other end to the flat sheet of a person's bed. When they come to bed and pull their sheets back . . . BANG!! This has been a favorite for a long time. However, there have been times when this trick has caused trouble. I have several sets of sheets for my bed that have black marks from being burned from the firecrackers. Once a sheet was set on fire but it was put out before anything serious happened. There have even been beds rigged so much that it sounds like machine gun fire when that person comes to bed. Sometimes it's amazing to think that this doesn't cause more fires.

We have had firefighters mysteriously wake up in other parts of the building, still in their beds. We had a firefighter who slept so soundly that we were able to dismantle his bed, carry him outside, and set his bed back up in the back of his pickup truck. Fortunately, it was not raining and it was warm outside. There have been gallons of shaving cream used in the night. A blob of shaving cream in the hand and a tickle on the upper lip has brought a lot of laughs.

Once, in a station with ceiling fans in the kitchen, a firefighter from another station snuck in during a class and poured flour on the blades of both fans. That afternoon, the cook turned to one of the other firefighters and asked him to turn the fans on. As soon as they came on, it looked like a bomb went off with all of the flour floating around. The firefighters swore revenge but to this day they have never found out who pulled the trick. The only reason I know is because the firefighter who did this couldn't resist telling me.

Then there was the Popsicle incident. We have a firefighter at our station, John Brenner, who had the habit of dropping his pants to his knees and telling people to kiss his skinny white ass. Everyone in the station was starting to tire of this little display. One day, he decided to do this to the wrong person. Bullet was sitting in the kitchen, eating a Popsicle. Bullet said something John didn't like. When the John came over and dropped his pants and told Bullet to kiss his ass, Bullet took the Popsicle and stuck it up John's ass. John was jumping around and screaming with that Popsicle sticking out of his ass, begging people to pull it out. Even if anyone had wanted to help him, they couldn't because they were laughing too hard. I'm not sure how he got it out, but since then he has not dropped his pants for an ass kissing. This has become a legendary story in our department.

Firefighter humor can, in the eyes of others, be a little on the cruel side. Calling burned bodies crispy critters can seem heartless. It's not that we don't care. If we never saw another burned body ever again it would be too soon. Calling drowning victims floaters may seem cruel, too. But people have to realize that if we dwell on the bodies or let ourselves think of them as being human, it would create even more stress. This job already has a lot of stress without adding more.

To look at a man who has died in a car wreck and then calling that man a dumbass may seem to be cruel to the public. But they don't see the empty beer cans or the half-empty fifth of liquor or the drug paraphernalia in the back seat. After seeing what can happen to sober people, they would understand why we would call a man a dumb-ass for intentionally going out and doing that to himself.

That's one of the reasons we play so many tricks on each other when we're in quarters. It relieves a lot of the stress we have from calls. It's one way of relieving the stress. We do talk about the bad calls when we get back, and that helps. We talk about the good ones, too. But it is always more fun to find a humorous way to relieve the stress.

One thing that I once got in trouble over was when a fellow firefighter tried to commit suicide by sitting in his car in the garage of his house. After an extended vacation in a sanitarium, he came back to work. Someone - and I am not saying who, except that it was not me - placed a muffler on his bed. Attached to the muffler was the mask off a bag-valve mask. Also attached to the muffler was a label with the words "Suicide Starter Kit." I had nothing to do with the joke but I did like it. The firefighter in question was an ass hole to everybody he ever met, anyway.

We had another firefighter who thought of himself as a real lady-killer. We figured that the only way he could "kill" them was with laughter. Anyway, while he was in training, a group of the rookies went to a bar and got hammered pretty hard. They had been drinking for several hours when the "ladies man" passed out in the middle of the dance floor. We came to the scene and checked him out. Then the paramedics arrived and did a more extensive evaluation of him. He regained consciousness and told the paramedics that his leg was killing him.

The paramedics determined that he needed to have his leg splinted. To do so, they had to cut his pants leg to get the splint on. He kept begging them not to cut his pants but they had to. When they reached his crotch, they found that "Mr. Ladies Man" had a nine-inch cucumber taped to his leg. We all rolled after seeing that. When he came back to training after healing, the training officer took the rookies out on a run through the trails at our training center. All of the rookies fell out with laughter when they looked in the trees along the trail and saw cucumbers hanging from them. About two weeks after getting out of training, "Ladies' Man" was caught by his captain masturbating in the shower. From then on, his nickname changed from "Ladies Man" to "Salad Shooter."

Chapter Four

Some of the toughest calls we run are calls involving children. I love to have them come to the station to see the trucks, but if we never run another call for children, it will not bother me one bit. I would just as soon fight a thousand fires as opposed to having to treat another child.

The first call I ever rode was for a child. A man had bought a brand new pickup truck and, to celebrate, he went out and got drunk. He made it home and decided to give his five children a treat and let them ride in the back of the truck. As he was going around a curve, he went too wide and caught the tires on the curb. The truck flipped several times and when we arrived on the scene we searched the scene for victims. The father was passed out in the truck but he didn't have a scratch on him. Four of the five children were scraped up but otherwise okay. However, the fifth child, the driver's seven-year-old son, was trapped with the bed of the truck across his stomach. When I reached the child, he wasn't breathing and had no pulse. The only way I could treat him was to crawl under the tailgate of the truck and treat him from there. Fortunately, there was enough of a gap that I could get under it.

I managed to take the jump bag with me and I was able to start CPR on him. After a minute of CPR I was able to bring him back. He had a pulse and was breathing normally. The problem was that the truck was not stabilized enough so that it could be lifted off the boy. He kept normal vitals for about three minutes and then crashed again. I started CPR again and managed to continue for about ten minutes until he came around again. This time he kept vitals for another two to three minutes but crashed again. The paramedics were busy intubating him and starting IVs, so I had to keep going with chest compressions. He kept coming around and then would crash again.

All in all, I did CPR on that kid for over 45 minutes. EMS transported the patient and I rode in with them still doing CPR. The last time I brought him around, he looked up at me and said, "Tell my mom I love her." Then he crashed for the last time. By this time I was exhausted and could not continue chest compressions. There were two paramedics in the back of that truck with me and they took over.

We left the hospital and went back to the station. I was still in training and just doing a ride along at the time. After I made it to my car, I sat there for twenty minutes and bawled my eyes out. The next day, I was going to play the piano for my sister's wedding. I was supposed to be at the church for the rehearsal when I was sitting in the car, but I had to compose myself before I could drive. I ended up arriving nearly an hour late and she and my mom were pissed. They jumped all over me about being late but I could not tell them the real reason. My sister is very tenderhearted and I knew it would upset her if I told her about the call. I just told her that something came up at work and I couldn't make it there any sooner. To this day, neither she nor my mother have any idea why I was late.

I have ridden calls where children were killed. I have carried their burned bodies out of their homes and I have carried their broken bodies from car wrecks. I have no idea why some parents cannot get it through their fat heads that their children need to be belted in their seats. Smaller children need to be in approved car seats. Sometimes I just want to smack the shit out of parents that do not make their children wear seat belts every time they are in the car. I have gotten a lot of parents in trouble after radioing communications and having them send a police officer to nail the dumb-shit parents who let their kids ride without seat belts. I've seen them packed in cars, riding on laps, and standing up. I even saw one parent who was letting her daughter, who should have been in a car seat, ride sitting straddle on the console of the car. I couldn't believe she could be that damned stupid. I not only radioed communications, I followed her until the police showed up. The mother was extremely pissed at me but I know that I did the right thing. It turns out that she lived near our station. Two days later, we pulled up beside her at a stoplight and I saw the child riding in a car seat. It seems that a $500.00 fine is a good way to get parents to use a car seat.

Calls involving the elderly are tough, too. When an old person is sick and we are rushing around trying to give them oxygen and start IVs and check vital signs and everything else, we have a tendency to forget that their spouse is right there watching everything that is going on. The person that they have been married to for forty or fifty years is lying on the floor in front of them, dying. Fortunately, I have Bullet with me. He is an expert at distracting spouses. He will get them to collect all of the patient's medicines and find out their medical history and doctor's name. Information that is needed by the hospital staff but information that we cannot get from the patient. Even though they know their spouse may die, it does give them something to do and it gives them a feeling that they are helping their spouse. And they are. We don't know where the patient's medications are, who their doctors are, or what their medical history is. We need that information. But the most important thing it does is it takes the spouse out of the room so they do not see what is going on. When you are doing CPR on a patient, they may throw up, they may shit all over the place, or both. There are bad sounds coming from their body. Sometimes we have to defibrillate them and their body will jump off the floor. The spouse doesn't need to see all that. The best thing that can happen on a call is when a friend is at the scene and can take the spouse out of the room.

I have pulled people out of fires when they were only seconds away from certain death. Some make it out unharmed and some are injured. Some survive their injuries and some don't. But we always do our best to find a victim. We have been successful in our searches and sometimes we can't find a victim. We have gone in search of victims that were in another location outside of the fire and in the process we have had some of our own people get hurt. This department has an excellent safety record. In the 104-year history of this department, there have only been two line-of-duty deaths. That is still two too many, but the point is that the number could be a lot higher. When a family gets out of the house after a fire has started, they should go to a specified spot. That way, everyone will be accounted for and there will be no mistakes.

We rode a fire at a furniture finishing business. When we arrived, we found everyone standing outside. The manager of the shop told us that everyone was accounted for except for a man named Jack. No one knew if Jack made it out or not. We searched that building three times, twice while the fire was burning and once after the fire was out. We turned over every square inch of that building but could not find Jack. About an hour after the fire was out, this man came down to our Battalion Chief and asked him if anyone was hurt. Chief Jones told him that one man was still unaccounted for and presumed dead. The man asked who it was and Chief Jones told him a man named Jack. Then Chief Jones was informed that the man talking to him was Jack. It seems that when the fire started, Jack decided he would just go home and come back later to see if he still had to go to work the next day. Jack then received a thirty-minute lecture from Chief Jones about the importance of reporting that you are okay. If he had reported to the manager at the location he was assigned, we could have put the fire out with a lot less damage. However, because we had to go look for him, that took two men away from fire suppression and put them in a search mode. As it is, the building was not that severely damaged and work would resume the next day for all employees. That is, all employees except for Jack who was fired on the spot.

It is also important to know who you have inside the house and where they are. However, you should also make it known WHAT is in the house. I rode a call to a house fire where a woman was standing outside screaming that her babies were still in the house. I asked her if she knew where they were and she told me that her babies are in the bedroom on the left at the end of the hall. We went in with two men on a hose line and two more searching for the babies. We went into the room and found out that her babies were two fully-grown Rottweillers. They were trained attack dogs and one jumped at me. I was able to block the dog from getting to my body by sticking my arm out and letting the dog latch on to my sleeve. The other dog got right behind the first and where ever the first dog moved, the backup dog would move. I knew I would not be able to carry her "babies" out so I just drug one dog with me while the other one followed. I got them outside and before I could say anything, my Battalion Chief grabbed me and put me onto another job. I never did get to tell her that she could take here "babies" and go to hell. I wouldn't have been nearly as upset as I was if I had known that we were looking for dogs.

One of the bad types of calls that you run are calls involving obese people. I'm not talking about people who are twenty or thirty pounds overweight nor women who are "retaining water." I'm talking about people who weigh 300 to 400 pounds or more. I have no idea why these people cannot live on the first floor. It always seems that they want to live at the top of a building with no elevator. In fact, that's one of the size-rating methods we have on this department. The bigger they are, the more floors you will have to walk up. And invariably they cannot walk to the ambulance under their own power. You have to carry them down. I've had to carry them as many as six floors. Even with six people carrying him out, it's not an easy job.

We had one gentleman that we rode on a regular basis who didn't live on the top floor of some high-rise building but who was much worse. The stench from the house would hit you out in the front yard. It always smelled like something had died in that house. It was so bad in the summer that we would have to wear our air-packs in when we rode his house. The first thing you would notice was his car. There was just enough room for him to sit in that car and stuffed all around him were loaves of French bread and bottles of wine. You could see where he would reach into the pile and tear off a hunk of bread to eat.

As we entered the house, the stench was unbearable. I must have rode by that house at least a hundred times, but I never got used to the smell. You walked in and after your eyes quit watering, you would see piles of shit in the floor. That's bad enough, but when you realized that this man did not have a pet, it made it even worse. When we would get to him, he would be lying naked in his leather recliner.

Apparently, he only wore clothes when he went out in public. He had everything wrong with him. Diabetes, bladder problems, cancer, you name it. Once, the paramedics lifted his belly up to do something and the skin on the underside of that roll of flab was rotting away. This had to be the nastiest man I have ever seen. He ended up dying about five years ago. To this day, his family has been trying to sell his house but that stench still hangs thick in the air. The fire department wouldn't even buy the house to burn because no one was sure about what would be released into the air if it did burn.

It's calls like these that really make you want to reconsider the line of work you're in. These are what we call shit calls. But it seems that when these shit calls come in, they are followed by simple calls or calls that make you realize that there is nothing else in this world that you would even think of doing.

Before, I said that one way of rating obese people is by figuring what floor they live on by how much they weigh. Another method we have of rating them is flushes. For every hundred pounds a person weighs, that's how many times they have to flush. For example, there were two brothers that lived together in our first call territory. One weighed in at 350 pounds, the other at 425. The first brother would have had to flush three times and the second brother four times. It's a little on the gross side, I know, but it is pretty accurate.

Chapter Five

We started our shift today with a little excitement. It seems that one of the guys off B-shift was fixing his breakfast in our old, beat up toaster. The toaster has a problem with the timer and it won't pop up anymore. You have to watch what you are cooking and, when it's ready, lift the lever on the side to get it out.

Tony Jester was cooking strawberry pop tarts and, after putting them in, forgot about them. Most people don't know it, but strawberry pop tarts get very flammable when they are over heated and the crust starts to burn. Everyone was out in the engine room when we heard the smoke detector going off inside the station. We ran in through the smoke and saw that the pop tarts had burned up in the toaster and had set the kitchen cabinets on fire.

I ran back out to the truck and told Bullet to start it up and charge the booster line. I got on the radio and said, "Engine 18 to fire communications. We have a fire at 1429 Elm Street. We need a battalion chief and fire investigator at this location."

Fire communications answered, "10-4 Engine 18. Confirming the address, did you say 1429 Elm Street?"

"10-4, fire communications. 1429 Elm Street. Station 18 is burning."

I ran back in with the booster line and put the cabinets out. It didn't take long to put that fire out but it did do extensive damage to the kitchen. Kitchen cabinets are not cheap and the fire burned four sets, two upper and two lower, beyond repair. You know, it's a strange feeling when you ride a fire call and never leave the station. I have never had to put a fire out in my own station before.

This is not the first time a fire station has caught fire. We had two fires in old Station 4 because the guys got a call and forgot to turn the stove off. Since Station 4 has been moved to a new station, there have been no fires there. I figure that either the jinx is over or the crews at Station 4 are getting tired of having the chief jump their asses after setting the station on fire.

Of course, we're not the only fire department that has had fires in their stations. There have been numerous fires started for one reason or another in stations all over the country. Some have started because of food on the stove. Some have started for other reasons. But it is strange to fight a fire in your own station.

After our fire, we had to clean the station and try to get the smell out. We had several people come by to ream us out for the fire but every time they started in, I showed them the same piece of paper. Toasters are supposed to be supplied to each station by Headquarters. Every time someone brought the subject up, I gave them the paperwork that showed that in the past year we had requested a new toaster twenty-seven times. Of course, they would not admit that they were wrong, but it did take the heat off of us.

That afternoon, we rode a car wreck in a bad curve on one street near our station. A car and a minivan were traveling west when the minivan lost control and hit the car. Both vehicles slid straight off the road and down an embankment. The car hit a tree on the passenger side door followed by the minivan. When the van hit the car, the car went straight up. The van came to rest against the tree and the car came down on top of the van. When we arrived, we found out the van driver was not hurt but could not get out because all the doors were jammed. The driver of the car, on the other hand, was extremely critical.

The first thing we did was stabilize the minivan. We put chock blocks under each tire to keep it from rolling and cribbing on all four corners to keep it from rocking. Then we started packing cribbing between the car and them minivan. Once the cribbing was in place, we tied the car off to the tree on one side and to our pumper on the other. Then we found a good sturdy tree on each end of the car and tied it off that way, too. For the first time we were able to make physical contact with the victim in the car. The van driver would have to wait until we got the car off the top of his van.

   I climbed up on top of the van and I was able to get the glass out of the back window with my center punch. I crawled inside with the victim and checked her out. She was breathing and had a strong pulse. The thing that surprised me was that there was not a lot of blood on her. However, I knew that we would still need to hurry because it was very apparent that she had internal injuries. She had a severely decreased level of consciousness and even though she was breathing fine at the moment, you could hear the fluids in her lungs without having to use the stethoscopes. I yelled for them to bring up the oxygen and I put her on high-flow oxygen. I had another firefighter get in the car with me to stabilize her head and keep her airway open.

   The paramedics arrived and I climbed out so they could get in there and get the cervical collar on her and start the IVs. Once they told me it was okay to start cutting on the car, I took the jaws and started cutting the posts between the windows as another firefighter broke the glass out of the windows. Once I had all of the posts cut, we removed the top of the car and started working the short spine board in behind the victim. That is when she started coming around. Every time we tried to move her, she would scream in agony. It didn't matter where we touched her, every part of her body felt like Jell-O. I don't think she had a bone in her body that was not broken. The victim was not conscious enough to be of any help with her extrication, but she was definitely able to let us know when we were hurting her. We finally just had to move her as carefully as we could but accept the fact that she would scream no matter what we did.

We got the spine board in behind her and strapped her to it. Then we started to turn her sideways in the seat so that we could get her laid out on the long spine board. When we started to turn her in the seat, we realized that her legs were broken so bad that they were not moving with the rest of her body. We then had to get the jaws back up there and pop the driver's-side door open. After we got the door open, we were able to get to her feet and lift her legs with the rest of her body. We slid her onto the long spine board and strapped her down to it.

We slid the spine board into a Stokes basket and strapped her in. We had to use the basket because she was 300 feet from the road and it was up hill all the way to the ambulance. We tied a rope to each end of the Stokes basket and the men at the top started pulling her up the hill while one man at the bottom used his rope to help steer the basket. We had a man on each corner of the basket to carry her up the hill and we each had a rope tied around our waists to help us out. The problem was that the embankment was so steep that we stumbled quite a bit and the victim screamed all the way up the hill. The embankment was so steep that I was surprised as hell that neither vehicle rolled over.

We got the victim to the top of the hill and loaded her in the ambulance. The paramedics both got in the back and one of our guys drove to the hospital. We were standing there watching the ambulance drive off and Bullet told me that we did a good job. I walked over to the squad and got a drink of water and to get a smoke. I needed a break for a few minutes. We stood around talking for about five minutes when we heard a faint voice off in the distance. We all looked at each other and then it dawned on us all at the same time. We had forgotten about the guy still trapped in the minivan. That's when we all ran back down the hill and started to work on getting him out.

We removed the car from the top of his van and then used the jaws to pry the sliding door open so he could get out. I didn't want to pry any door with the car on top of the van for fear that the weight would collapse the roof. Once we got him out, he was very appreciative. He thanked everyone and just went on and on about how much he thought of firefighters. Then he told us that he was starting to think that we had forgotten him and wanted to know what took us so long to get back down there to him. I just said, "We didn't forget you. We just needed to put some things together to get you out. We could never have forgotten you. Because, without you this would have never have happened." The man understood and thanked us again. We watched as he started up the hill and then he stopped and turned around. He looked straight at me and all I said was, "Have a nice day, sir." I think he realized the sarcasm and started to say something and then thought better of it.

On the way back to the station, we stopped and got some groceries. While we were sitting in the parking lot, a man came up with his four sons. We found out that their names were Jason, Tommy, Billy, and Mark. They ranged in age from two to eleven. They wanted to see the fire truck and we were more than happy to show it to them. Each of the boys took a turn sitting in the driver's seat and then they wanted to see the lights. Bullet hit the lights for them, blew the horn, hit the siren, and then started to show them some other things on the truck. These kids were fascinated by everything they saw. Then we let them try on the helmets and their father took their pictures. In all, we spent more than thirty minutes out there showing the boys the truck. I was hungry but I love dealing with kids so I didn't mind.

Once they left, we headed to the station. The guys on the other trucks were pissed that they hadn't eaten dinner yet and it was getting late. Chief Jones was sitting at the head of the table with an empty plate in front of him. He had a knife in one hand, a fork in the other, and he was staring at his plate. He looked at me and said, "Dalton. Do you see this plate?"

I said, "Yeah, Chief, I see it."

"Do you notice anything wrong with this plate?"

"Are you referring to the fact that it's empty?"

"Oh yes. Yes I am. You know how I get when I'm hungry. And right now I'm hungry enough to eat a horse. So you better get a jump on supper."

Pointing at his gut, I said, "Well, Chief, I'd say that you can stand to wait a little longer on supper."

Chief Jones stood up and looked me in the eye and said, "Boy, are you saying I'm fat?"

"Not at all, Chief. But I have noticed that the ass on your shadow weighs about forty pounds."

Everybody in the room got a good laugh out of that one. Even Chief Jones. He likes to act tough but he's one of the best people a person could know. Chief Ray Jones is the type of person that is not afraid to tell you that you screwed up, but he will also let things go after he has chewed you out. Some chiefs on this department can definitely hold a grudge. I've known of some that would go out their way to get at a man. We had one chief that got upset when a hose load came off the back of the truck as they were pulling out of the station. The driver stopped the truck and they loaded the hose back on, correctly this time. The shift before them had loaded the truck wrong and the load came off. This chief showed up while they were loading the truck, and after hearing what happened, laid into the driver. It was not his fault but, since the chief could not chew out the shift that messed up the hose load, he decided to give the driver hell about it. When the driver informed him that he was not at fault, the battalion chief told him that if he kept on back talking him that they would be taking a trip to the fire chief's office. Not to be outdone, the driver said to the chief, "Well, goddamn it, we'll go right now!" He proceeded to walk over the chief's car and get in. He honked the horn and yelled to the chief to get his "ass in the car and let's get this thing over with." The battalion chief made him get out of the car and decided not to go to the fire chief with this. But he never forgot about the incident, either. He gave that driver so much hell that he ended up transferring to another shift to get away from him.

That's how a lot of battalion chiefs are, but Chief Jones is different. It took him a long time to get to the rank of battalion chief. Probably because he isn't a suck-up. Chief Jones isn't one to go along with the crowd. If he thinks something is wrong, he is going to say so. The other chiefs don't like that quality in him. But one thing that can definitely be said about Chief Ray Jones is that he is a fair man. He doesn't play favorites and he will not let anyone suck up to him. If they do, he'll be the first one to give them hell about it. If you screw up, he will not chew you out and embarrass you in front of the other men. He likes to deal with people on a one-to-one basis. If the entire company screws up, he'll chew us all out. But he will not do it in front of another company. And, once he is done chewing you out, the subject is dropped.

Chief Jones has one other quality about him that I have never seen in another chief. He will admit it when he is wrong and if a mistake was made, he will make it right. If he comes in and chews out a man and finds out that shouldn't have happened, he will come and apologize. No other chief I know of will do that. If he chews out the wrong person, he will apologize and then go chew out the right one.

I finally got the chief's dinner ready and he inhaled it, as usual. Saturday is steak night at the fire station and I have never seen anyone that can put a hurting on a steak like Chief Jones. I served him first, and by the time I sat down, he had finished eating and was gnawing on a toothpick. It's a shame that he ate it so fast, though. I think he would have really liked how the steak tasted.

We were finishing up with the dishes when the tones went off. A reported house fire on Main Street. When we arrived on the scene there were three people standing around a man lying in the middle of the street. We found out that the house was not burning but the detached garage was. The men on the squad started treating the man in the street as we pulled a line and hit the fire. There wasn't much in the garage except for a few tools, a fifty-five gallon drum, and various car parts. The car that he was working on was out in the driveway.

We were able to knock the fire out in about ten minutes and we started looking for a cause. At that point, we thought the man was overcome with smoke and had walked down the driveway to the street. But then we noticed that the garage itself was sitting about three feet to the right of its foundation. That's when Chief Jones told us that the neighbors reported hearing an explosion. The paramedics were getting ready to leave with the man when Chief Jones asked them if he could talk to him. He got in the back of the truck and stayed in there for five minutes or so. He came out and when the doors on the ambulance shut, he burst into laughter.

The car that he had been working on had been running hot for a while and the motor was also running rough. The man had boiled out his radiator a couple of weeks ago and that stopped his overheating problem. He found out the day before that the reason the motor ran so rough was due to trash in the fuel lines. He figured since boiling out his radiator worked so well, why not try the same thing with his gas tank. It took about five minutes for the fumes remaining in the tank to explode. The reason the man was in the middle of the street was because that is where he landed. Had it not been for the fact that he had his garage door open, his injuries could have been a lot worse. It's better to roll down your driveway than to be blown through a door. We found out later that the man's injuries were minor and that he would be going home the next day. The doctor told the paramedics that he was going to keep the man overnight so that he would not end up hurting himself even more.

I decided to go to bed early that night which is usually not the best thing to do. It seems that if I stay up past midnight, we don't get many calls at night. But the minute I try to go to bed early, we end up getting slammed. And this night proved my theory right. There were no serious calls, but just enough of them to screw up a good night's sleep. We went out twice.

The first call was for a fluid spill at a vehicle accident. No one was hurt and there was no fire, but the radiator on one of the cars was punctured and the fluid was headed towards a street drain. It used to be that we would simply hose the street down. But in this day of environmental awareness, now they send the fire department out to put down absorbent, which is basically kitty litter, to soak up the fluids. Then the street department comes by and gets rid of the absorbent . . . . usually by sweeping it down the storm drains.

The second call was a real moron. He had been at his house drinking all evening and then decided at 2:30 in the morning that he would drive to the store to get more beer. He ended up driving off the road, sliding down a hill and slamming into a tree. When we arrived, he was standing on the side of the road looking at his car. I asked him if he was all right and said he was. I asked him what caused him to wreck his car.

He told me, "I don't know. I was driving to the store to get some more beer when all of a sudden I slid off the road and hit that tree."

The first obvious question, "How many beers have you already had?"

"I don't know. About nine, maybe ten."

Then I asked the second obvious question, "You do know that in this state you cannot buy beer after 2:00 AM?"

"You mean I wrecked my car for nothing?!?!"

"Well, look on the bright side. Nobody got hurt and you were able to get another drunk driver off the road." He didn't have a chance to say anything else. A state trooper took him into custody.

We waited on the tow truck and, of course, they had to get one from out of town, so we had to wait over an hour. I would have preferred to have gone back to the station and go to bed but we had to wait so that we could make sure nothing happened when they towed the car off. Of course, nothing happened, but if we had left we probably would have been called back out because the car caught fire or something.

Once the car was out and on it's way to the junkyard, we headed home. I went to bed for the third time that night and I was able to sleep until communications set off the wake-up tones at 7:00. I could have used another couple of hours of sleep but I figured I could get that at home just as easily as I could at work. And I could crawl into bed with my wife, Teri, and sleep with her instead of sleeping all alone. Besides, she doesn't snore like the guys in the dorm. The only problem with sleeping at home is that not only do I sleep with my wife, but I also have to sleep with three dogs. They are the best dogs in the world and I love them to death, but I do wish I could get them to sleep pointed from headboard to footboard instead of trying to sleep from side to side on the bed. Sometimes, it's hard to get up because they lay on top of the covers and you are pinned in the bed. If I could just convince Chief Jones that I can't get out of bed because I'm pinned in, I could use that to my advantage. But I don't think he would buy it.

I gladly went home and crawled into bed with my wife and our three dogs. I'm just happy that they were the only ones in the bed. On several occasions, I have come home, ready to go to be and found that there was no room. Not only were my wife and our dogs in the bed, but our son or daughter or both were in there, too. No room in the inn for me.

About twenty minutes after laying down with my wife, she woke me up and asked me if I wanted to go to church with her. I figured that I would go so I could be with the family. Besides, in this line of work, you need all the help you can get.

She asked me, "Are you sure you want to go?"

"Yeah, baby. I'm sure."

"I just know that sometimes you need to get some sleep."

"Hell, I'll get all the sleep I need when I'm dead."

"So how was your night?"

"It went."

"I know what that means."

"Huh?"

"Every time you say that, I know that you had a rough one or you didn't get enough sleep."

"Well, I'll be all right. Are Joey and Nikki up?"

"Yeah. They've been up for a few minutes. I need to get their breakfast."

"Okay. I'm going to shave and shower. I'll be ready in a few minutes."

"Oh, by the way…" She walked over, gave me a hug and a nice long kiss. We've been married for fifteen years and I still love to kiss that woman. She's a great kisser, too. And after having two children, she still has an excellent body. I wish she would take a compliment once in awhile. I tell her she looks good and she tells me I need new glasses. I just wish that once she would accept what I say and say thank you. I really do believe that my wife is the most beautiful woman in the world. If only wish that she would let me say so.

After getting ready, I choked down breakfast and we left for church. I had to fight to stay awake, but I managed. I'm not exactly sure what all was said in the sermon, but Reverend Gray was really rocking and rolling. All I could think about was that I had to stay awake. I fell asleep in church one time before and pissed Teri off. I not only snored a couple of times but I also farted. I couldn't help it. I was asleep and I didn't know what I was doing. It just slipped out. My wife didn't find the humor in it, though and didn't talk to me for the rest of the day. She couldn't believe that I could do such a crude thing in church. She was "so embarrassed." I just let it slide and went on about my day. I did, however, change "Chili Night Saturdays" to "Steak Night Saturdays." In fact, that's the last time I have had chili on a Saturday when I may be going to church the next day.

After the service ended, the congregation was filing out of the church, shaking hands with Reverend Gray and telling him what a good service he performed that day. I made my way over to him and as I shook his hand he said, "Dalton, we haven't seen you in here in awhile."

I said, "Yeah, well I have been working a lot on Sundays lately and I haven't been able to make it here."

Before either one of us could say anything else, one of the old biddies of the church asked, "And just what kind of job do you have that would take you away from worshipping God?"

I turned to her, looked her straight in the eye, and said, "I'm and arsonist. I burn things up for a living."

I thought Reverend Gray was going to fall out in the floor. Before I left, I went back to Reverend Gray and asked him if he ever told her the truth. He said, "No! That one was too good to let go. I thought she was going to have a heart attack when she heard that."

"Well it's a good thing that she didn't. I'm off duty."

Reverend Gray laughed and walked away. I got in the Jeep and asked Teri where she wanted to go. She said she didn't care where we went but she wasn't cooking anything today. So I turned around and asked the kids, "What do you want for lunch?"

Nikki said, "It doesn't matter to me."

Joey said, "Pizza!"

I think if it were possible, Joey would live off of pizza. Of course, he gets it honestly. I am the pizza king. I can eat it any time of the day. And it's healthy. You have all four food groups: the crust for the bread and cereal group; the tomato sauce for the fruit and vegetable group; the cheese for the dairy group; and the toppings for the meat group. See all four basic food groups and a little grease to add flavor. What else could you ask for? And pizza is not just for breakfast any more.

We pigged out on pizza and headed back home. I went and took a nap while Teri helped Nikki rearrange her room and Joey played on the computer. When I got up, I went out and threw the football with Joey until it was time for dinner. We went in and I grilled some burgers. We had a picnic on the patio.

The thing I love about Sundays is the fact that I get to spend time with my family. Nikki, my beautiful daughter, who is nine. Joey, my wonderful son, who is ten. Teri, my lovely wife, who would kill me if I told you her age. Then there is me, Dalton, the loving husband and father, who is old enough to know better but still too young to care. And, let's not forget the dogs - Gizmo, Mojo, and Jake. Together, we are the James family. I have the best family in the world. I know that there are those who would dispute that, but you will never convince me otherwise.

My daughter, Nikki, is the creative one. She loves to make things, draw and do things to make the house look prettier. She loves all forms of art; music, pictures, etc. She just turned nine on the twenty-third of this month. She's getting into makeup and hair, as well. She can come up some pretty wild stuff but no matter what she comes up with, she is still cute as hell.

My son, Joey, is the curious one. He loves to play on the computer. He also loves to take things apart and put them back together. I just pray that one day he doesn't decide to combine those hobbies on my computer. He loves to do things to make money. He will make deals with his grandfather to get money. Such as the time he made a deal with him that for every wheelbarrow load of leaves he hauled from a spot in the yard to the burn pile out back, he would get a certain amount of money. He even had different rates for the front yard and the back yard and he carried a notepad around to keep track of his loads. I guess it doesn't hurt that his other grandfather drives a truck for a living.

My wife, Teri, is the one who runs this house. I know that every man out there thinks they are the kings of their castle, but it's the lady of the house who is really in charge. Think about it. Most men don't know where half the stuff in their house is. They can spend two hours looking for something they need and then ask their wife. She can stick her hand in the pile he has created on the floor and without looking pull it out. And what does she do next? She makes you clean your mess up. I'm a little smarter than most men are. Now, I just ask Teri in advance if she knows where something is. She usually does and it avoids a mess. Less work for me.

Teri is the best person in the world. She is smart, beautiful, sexy, has a good sense of humor, and a good head on her shoulders. And she has great taste. She married me. I have seen her go through more in her life than most women can even imagine and still come out on top. She is my heart. My soul mate. She is the one thing I have in my life that keeps me from going nuts. She keeps me straight. I love her more than life itself.

I love my entire family more than life itself. Maybe that's why I am careful on the job. Maybe they are the reason that I don't get careless. I know that I have to be with them and that keeps me from trying to be "Joe Hero" on the job. Don't get me wrong. I will do as much as I can do to help people when I am on the job, but keeping my family in mind when I go into a fire helps me find a way to work safely. My family is what anchors me to the ground and keeps me from losing my head.

Now comes the best part of Sundays in the Dalton household. Every Sunday evening, when the dinner dishes are done, we sit down and play games. Of course, it doesn't hurt that there usually isn't anything to watch on TV. This particular evening, the kids wanted to play poker. They have been playing on the computer and now they wanted to try their hand at real poker. There wasn't any money involved, of course, but we still have fun playing. And I taught them a very valuable lesson about winning and losing. They were both kind of cocky about their poker abilities after beating the computer and I was able to show them how to lose with dignity. The cards just weren't with me this time. Of course, they're usually not with me, which is why I don't ordinarily play poker.

Once the kids were in bed, Teri and I decided to hit the bed. I love cuddling with my wife. She gets cold very easily and I get to warm her up. It's a pretty good deal. And I love nothing more than when she lays her head on my chest and I get to hold her. And after a while, our puppy, Jake, figures that's enough cuddling time and wedges himself between us.

The sleeping arrangements are set in stone. On the left side of the bed, we have Mojo, then me, then Jake, then Teri, and on the right side we have Gizmo. All I know is that the king sized bed we have needs to be enlarged. But I don't mind it. I always sleep well at home.

The next day was uneventful enough. I went to my part-time job and replaced the upholstery on a couple of couches. I came home in time for dinner with the family and then went and threw the football with Joey and Nikki. Nikki may be into makeup and hair but she can snag a pass like nobody's business. Joey is a pretty good receiver, too. He's got good speed and he can catch well. Now if I could just get him to stop doing that touchdown dance of his. I know that most TD dances are goofy looking, but this one is the goofiest of all. But he likes doing it so I don't say anything.

After playing ball, I had to go get things ready for work the next day. Then I put the kids to bed and joined Teri in our bed. To do that, I had to crawl up from the foot of the bed to get to my spot. I leaned over Jake and kissed my lovely bride goodnight. And as I drifted off to sleep, I told her that I loved her.

Chapter Six

The day started off uneventful enough. We went to an EMT continuing education class for three hours and went back to the station for lunch. That afternoon we hung around the station cleaning and mowing the yard. There wasn't much going on. I guess everybody was inside enjoying the air conditioning because it was hotter than eighteen yards of hell outside. We ended up mowing the yard in shifts.

When the yard was done, we had a short class on a new nozzle we were getting. Nothing too hard to learn but it will be a good nozzle to have. And of course, we had to piss off the training chief, but we don't like him anyway. He has a tendency to get long winded and when we have been over the information three or four times, it's time to move on to something else.

He kept saying, "Now, you guys need to pay attention. You may need this information in the future." Of course, we all knew that he didn't know his ass from a hole in the ground, so we didn't listen.

It would have been nice if the night was as slow as the day was. For all of it's boredom, at least we stayed in the house and in the air conditioning. That night turned out to be a lot different. The first two calls we rode were simple enough. Two med calls, one for chest pains, the other for shortness of breath. We arrived at the same time as the paramedics so we didn't have to do a lot on the calls except load the patient in the truck. The last call was different.

At 3:30, we received a call to a reported house fire. Reported fires are worse than alarm calls because they are usually real. Alarms are commonly false alarms, but reported fires are, for the most part, serious. When we arrived on the scene, the house was fully involved. Fire was coming out every window in the front and all but one in the rear of the house. At 3:30 AM, people are usually at home in bed. We tried to enter the house a couple of times but we could only get into the house about ten feet. When we arrived, we didn't see any family members outside the house. We were afraid that the family was killed in the fire.

We started hitting the fire from the outside. We had four hose lines down and were hitting it through the front door and three windows. We were able to make some headway into the house and get the fire knocked down in the living room. There was a large bookcase in the living room that the owners were using as a room divider. Instead of being up against the wall, it was perpendicular to the wall. We had to move around it to hit the fire burning on the other side. As I was making my way toward the hot spot, I stepped onto an air-conditioning vent in the middle of the floor and, because the floor was weakened by the fire, I went through up to my knee. The leg of my bunker pants was pulled up and hot embers fell into my boot. I was able to pull my foot free of the hole but I had to get another firefighter to fill my boot with water to put the embers out.

I hobbled out the front door and told Chief Jones what had happened. EMS was standing by on the scene and Chief Jones sent me over to their truck so they could check me out. My leg and foot were red, but nothing serious. They told me that I could go back to work if I wanted to or they would take me to the hospital to be checked out further. Because I didn't see anything wrong and it only felt like a sunburn, I went back to work.

We started concentrating our efforts toward the bedrooms. The first bedroom we came to was totally gutted and the drywall had collapsed so we could see straight in. The bed was burned down to the frame and we were fortunate to find nobody there. We checked all around the room and in the closet and found no victims. We saw a set of bongos that were burned beyond repair and two burned out amplifiers.

We made our way down the hall to a closet. When children are scared and unsure of what they should do, they commonly crawl into a closet to hide where they think it is safe. We dug through everything on the floor of the closet and it was clear, too. From that vantage we were able to direct our streams into the bedroom on the front of the house. The fire was easily knocked down in that room because there wasn't much left to burn. We had two lines in the hallway and I directed two firefighters into the front room to do a search while we moved to the other side of the hall and started hitting the master bedroom.

The crew in the front bedroom gave an all clear. They started hitting some hot spots in the closet and where the bunk beds were. They were burned down to their frames as well and there was a pile of burning debris on the floor under the frames. Once those fires were put out, the crew started overhauling the room. There wasn't much to do in the overhaul. The ceiling and roof above it were gone and the walls were washed away with our hose streams. The room was pretty well burned out so there was nothing to salvage except toys in the closet. The crew started handing toys out through the front window as they dug through the closet, checking for any remaining fire.

In the meantime, another firefighter and I made our way into the back bedroom. With the exception of this room, the fire was out. We hit the seat of the fire and drowned it out pretty quickly. Once the steam lifted, you could see there was nothing left of that room either. And, fortunately there were no victims. We figured the family was out of town and had no idea their house burned down. The dresser had stayed intact enough to be carried out but it was otherwise useless. The clothes inside weren't burned but they were heavily damaged by smoke. Then we tackled the closet. It was full of papers, children's toys, and recording contracts. Apparently someone in the house worked for a local recording studio and stored the contracts in their bedroom closet.

We dug through that closet for an hour and a half looking for fire extension and carrying out what wasn't burned or destroyed by the water. We didn't find anymore fire, and we gave a loss stopped on the house but that didn't mean much at that point. The entire roof of the house was gone. The whole house was gutted. Everything in the kitchen, living room, and first bedroom was burned beyond recognition. With the exception of a few clothes and some smoked-up toys, the only thing the family had left was the clothes on their backs.

As the sun came up, we were taking a break. As we had coffee and donuts, we were looking at the damage done to the house. All in all, it was a smooth operation. The house was pretty much gone, but no one got hurt and the family, as far as we knew, was fine. We stood around until the fire inspectors came and then went back to work helping them look for the cause.

A little while later, the family came home. We found out that the mother had some form of cancer and she was having a bad night. The father decided to take her to the hospital at around 1:00 am. Being that late at night, he didn't want to wake anyone up to come and look after the children so he decided to take the three of them with him to the hospital. The daughter, who is thirteen years old, had broken up with her boyfriend and was burning his love letters and a jewelry box he gave her in the fireplace. She told her father that it was her way of "reaching closure in the relationship." The papers were still burning in the fireplace when they left. They didn't have a screen covering the opening and a couple of sparks must have popped out and lit the carpet on fire. I felt so sorry for the family. After going through a fight with cancer, now they have lost their home and nearly everything they owned.

We were carrying things out of the house and placing them in two piles. One was a salvage pile the other was a junk pile. The junk pile was growing by the second as the salvage pile grew very slowly. It was 8:00 am and crews were starting to arrive to replace us. We changed shifts on the scene and we rode back to the station in one of our Suburbans. Before we could get in the Suburban we had to take our bunker gear off. As I took mine off, I could see that my leg was worse than I had previously thought. My leg had an eight-inch blister down the outside, along the curve of the calf muscle. There was another blister that had busted on the inside of my leg just below the knee. A third blister was at the curve between the top of my foot and where it curves up to the shin. Most of the blisters had burst and soot and dirt from the embers. It looked pretty nasty.

We got back to the station and I took a shower to see if I could get it cleaned up. The black would not come off so I decided to go to the city nurse to see if there was anything that I needed to do to keep from getting an infection. She cleaned it as much as she could with a liquid antiseptic. I was dreading going there because I thought the burns would have to be debrided. Debriding involves taking a stiff bristle brush and scrubbing the area, something I was not looking forward to. Fortunately, she determined that the burns were only second degree and I would not need debridement. I was definitely thankful for that.

I ended up the shift filling out an incident report and had to tell several people what happened. I think it's worse to have to keep telling the same story over and over to every supervisor in the department than to get burn. One thing's for sure, getting burned is a hell of a lot easier. I definitely would have preferred to have a slow, boring shift.

One of the problems with working a part-time job on your days off is when you are up all night working the night before. When I finally got to my part-time job after filling out all of the paper work, I was two and a half hours late. Of course, my boss, Rick Jackson, knows that when I am late, I have a good reason. He understands and he always tells me that if I need time off because of my fire department job all I have to do is call him. But I had a couple of projects that I have been working on for a week that I needed to finish up so I decided to go on in. Besides, it wasn't too much work and I was feeling all right.

I went and told Rick why I was late and of course I had to tell him, his secretary and a couple of other ladies in the office the whole story. I spent another hour talking to them and when I got downstairs to my area of the shop, the supervisor there, Dick Randolph, jumped my case. I know that he is usually an ass hole and he has his ways, but I was in no mood to hear any of it. I told him not to start with me.

"You damned firemen always think you're special. Just because you work at a dangerous job you think you get special treatment here. Well I'll tell you one thing, you ain't getting shit from me."

"Look, I told you not to start with me today. I was up all night, I was burned, and I haven't had a good day. Now back off!"

"You think I give a shit?"

"Look, god damn it! If you don't back off, I'm gonna rip your head off and shit down your neck! "

He started to say something else when Rick walked in and wanted to know what the hell was going on. I turned to Rick and said, "Nothing. Just that after the night I had last night this ass hole decides to live up to his name."

Rick told me not to worry about it and go on with my work. Then he told Dick to come with him. I don't know what he said to him, but when Dick came back in, he was pissed. He gave me one of the best go to hell looks I have ever received. I didn't give a shit. I have never liked him and I didn't care one way or another whether he was pissed at me or not. All I knew was that I was not going to take any of his shit anymore.

I finished up my two projects and decided to go on home. Of course, when I left early, Dick had a smart comment for me as I walked out the door. I didn't really hear what he said so I just ignored him. I don't think he liked me ignoring him. I did, however, notice his truck parked two spaces from my Jeep. The temptation in me started growing and I couldn't resist doing something. I walked over to the house behind the parking lot. The owner has two St. Bernards on zip lines in the yard. I took a bag out of their trashcan and walked over to the dogs. They are friendly and I have often gone over there a lot to pet them. I took the bag and stuck my hand in it. I reached down and picked up three big piles of dog shit. I then turned the bag wrong side out and walked back to Dick's truck. He never locks it so I opened the door and dumped the bag up under his dashboard. I made sure to roll his windows up real tight and I closed the door. Then I threw the bag in the back of his truck.

I left and headed over the YMCA to pick Joey and Nikki up. They spend the days there in the summer and they play games, swim, do crafts - whatever the counselors come up with. Joey was working in the wood shop and had to show me what he had made. It was a towel rack for their bathroom. He did a damned good job. Then we went over to the swimming pool to get Nikki. I asked one of the counselors where she was and she told me that Nikki had seen me come in and decided to go get dressed. She came out a couple of minutes later and was ready to go.

We went on home and decided to plant ourselves in front of the TV. I knew that Teri would be home soon and I knew that she would be tired from working in the hospital ER so I decided to order a pizza. When she came home, the pizza guy pulled in behind her. Then I had to tell her about getting burned. She was a little pissed that I didn't tell her sooner but she let that slide. I still had the bandage on my leg but it was time to change it. I took it off and showed her the burns. Teri got a little upset about the whole thing. She worries about me when I am at work, but I still cannot give the job up. And she wouldn't think of asking me to give it up. She knows that I love my job and she wouldn't ask me to give it up for anything. And for that, I love her.

She helped me put the ointment the nurse gave me on my burn and dress the wound. It didn't take long and she helped me make it feel better. We then ate dinner and I decided to call it a night. I had to work my part-time job tomorrow and didn't have to get up as early but I was tired and needed the sleep.

The next day, I found out that Dick found his little surprise. He didn't say anything to me about it but I know he suspected me. Of course, he was an ass hole to everybody that worked for him so he couldn't definitely pin it on me. I didn't tell him that I did it but everybody there got a kick out of it. I also found out that he received a ticket on the way home for littering. The bag I threw in the back of his truck blew out on the way home and a cop pulled him over. And - you guessed it - he lived up to his name so the cop wouldn't let him out of the ticket. Oh, well.

The day went by pretty quick and I did the usual afternoon routine. I picked up the kids and went home and cooked supper. And of course, I had to remind my wife that she was a damned lucky woman to have a man like me. She asked, "What do you mean?"

"Who else do you know that has a husband that will pick up the kids and come home and cook you a damned fine meal like this?"

"Not many. I guess I am a damned lucky woman."

"Well, if you play your cards right, you might get even luckier tonight."

"But, what if I feel lucky enough right now?"

"Well then you'll be less lucky tonight." She gave me a puzzled look. "You'll see a grown man cry."

She laughed, "Well, we can't have that. I'll see what I can do."

With that, she grabbed my butt and gave it a squeeze. I called her a tease and she giggled. But I will say that when we went to bed, we were both DAMNED lucky!

Chapter Seven

The shift started off pretty slow today. We did station clean up and had a short class on streets. After lunch, there wasn't much to do until around 4:00. We got a call to go to 412 Elm Street, single company, non-emergency to assist the nurses. We knew exactly what was in store for us.

Four-twelve Elm Street is an assisted living home for disabled veterans. The men that live there are capable of taking care of themselves for the most part, but they do require a little assistance every now and then. The only time they call out a single company to come non-emergency is to assist with our friend, "Filthy Herman." Herman is a computer genius and is one of the nicest men you could ever meet. He lost both of his legs to a land mine in Vietnam. He came home and, rather than feel sorry for himself all of the time, decided to go to school. He got into computer systems and completely loves his job. In fact, he was the one who designed the computer system for the city of Mannington.

His only problem is with alcohol. He only drinks once or twice a month, but when he does, he totally changes. When he is drunk, he likes to roll his chair out on the front porch and strip naked. Then, as people walk by, he waves his penis at them and says hello. He is a gentleman but he is completely naked and waving his boner at people. The only time I have ever seen anyone get really mad at him is when he told the mayor that his dick touched the floor when he stood up and the only reason she didn't like him is because hers didn't. For some reason, she took offense to his statement.

I was the one who gave Herman his nickname. His real name is Herman, and the first time I met him, I had just finished reading Joseph Wambaugh's "The Choirboys." I named Herman after a character in the book that was a lot like him.

When we get called out there, the women that work in the house need us to get Herman dressed. It's not that he wouldn't get dressed, it's just that he keeps moving so much that they cannot dress him. They call us out and we have to do it. He is nice as can be to us but he moves around so much that it's like dressing a fifty-year-old, 140-pound baby. When we are done dressing him, it's like we have just gone ten rounds with the world champ.

The only other time I have ever seen anyone get mad at Herman's antics was last Christmas. He wanted to show that he was in the spirit of the season so he decided to decorate his "little friend." Somewhere he got a small elf costume and put it on his penis. As people walked by the house, he was sitting on the porch waving his "elf" at them and singing, "Jingle balls, jingle balls, jingle my balls all day." A couple of women took offense to this and called the cops. They came, told him to sober up, and then we dressed him.

When we got back to the station we had to hear it from the other guys. The same old comments, "Did you get lucky?" "Did you cop a feel?" The usual. We hear it from the other guys but they get to hear it from us if they ride him and we don't. It's the usual thing. When you get regular calls, what we call "frequent flyers," you get the usual questions. One thing is for sure, Herman is one of the better frequent flyers than the others. One is an ulcer patient and every time we ride him, he is on the toilet painting the walls with a stench that would gag a maggot. Another is an old lady who is a hypochondriac. She's lonely and wants company. We have others, but those are the more colorful ones we ride.

We sat around the table after supper, talking about things that were going on at Headquarters. We had a new fire chief and a new administration. People were bitching about some that were promoted and raving about others. The new chief sent out a memo that morning about a new pay scale system that he was going to propose at the next city council meeting. It was a good plan but we knew that it would never pass because city council has never been on our side.

If there is one constant thing with the fire department, it is that city council will never have any respect for the firefighters. They have never felt that we are worth the money we get paid. They feel that we are only in this job for one thing, money. If we wanted to do nothing but make money, why in the hell would we be working for the fire department. What we want is a fair wage for a job well done. When a fire starts, they think that anyone can put it out. But who are they going to call, the garbage man? It takes a different kind of person to go into a burning building and battle the beast. The people on the city council don't realize that. They don't realize that if their house catches on fire, they are going to get out and call the fire department. There are a lot of people out there that think they can do this job, but they realize that it is a lot tougher than they think. It's hot. It's dark. It can scare the hell out of you.

When you go into a house fire, it's not like watching it on television or in the movies. There is usually so much smoke that you cannot see your hand just inches from your face. You don't see that in movies. There aren't many people who would want to sit there for two or three hours watching a blank screen. I know I wouldn't.

The movies don't allow you to feel the heat in a fire, either. Temperatures can quickly reach 400 degrees or more in a house fire. That kind of heat can kill you in seconds. Extreme temperatures are very deadly. One breath of super-heated air can blister your entire respiratory tract.

At one city council meeting, one of the councilmen, a rather large man, claimed that being a firefighter was not as physically demanding as we were making out. I was ready for this because he had already made the same statement to me before. I brought out an air-pack and face piece and asked him to prove his point. There were a lot of voters in the audience so he couldn't back down. As we helped him get the air-pack on, he was complaining about how much it weighed. I just informed him that he should be thankful that he wasn't wearing turnout gear too because that would double the weight.

After we got him in the air-pack, we had him go out the main doors of the council chambers and get on his hands and knees. He was a little leery of getting in that position and Bullet, from the back of the group, said, "Hell, you put us in that position every payday." The councilman ignored Bullet's remark and got into position. The council chamber is a center room with a hallway going all the way around it. We had him crawl on his hands and knees through that hallway, a distance of approximately 400 feet. We then let him stand up and walk back into the council chambers and got him out of the air-pack.

Another councilman asked him how he felt. Between gasps for air, he said that it was a lot tougher than he thought. I then pointed out that he had no obstacles to overcome and could see everywhere he went. I also pointed out that moving around with only a fifty pound air-pack is a lot different than doing the same thing with a fifty-pound air-pack and fifty pounds of turnout gear. I do believe we made our point because he never made those smart-assed comments again.

We moved into the day room so that we could watch the ball game. It wasn't much of a game but we had fun teasing Bullet because he picked a loser to root for. We had a good time even though there wasn't much to watch.

We were throwing things at him and every time the other team scored we would rub it in. Then the tones went off and the party was over. It came in as a vehicle accident with injuries.

When we got on the scene, we found out that a dump truck had run into the side of a church. The cab of the truck had completely gone through the wall and the bed of the truck was jammed under the top. We surveyed the scene from the outside first and found that there was no immediate danger. We then went in and checked the fuel tanks first to make sure they were not leaking and to make sure the motor wasn't running. The only leak was from the radiator and the motor was off. We saw a man standing next to the truck and I asked his name.

"I'm Frank Taylor. I'm the minister here."

"Were you in the building when the truck hit it?"

"No. I was at home, next door. I heard the crash and came out to find this. I can't believe this. What are we going to do?"

"Well, let's get this truck out of here and we'll see how bad the wall is damaged."

"I think the driver's dead. He hasn't moved since I got here."

"We'll check him out, but you need to go outside. We don't want you getting hurt, too."

The minister left and we started pulling debris off of the truck to get to the door. Fortunately, everything was loose so we didn't have to cut anything. We checked on the beam at the top of the wall to make sure it wasn't impinged or cracked. We found one small crack and started shoring up the wall so it wouldn't come down on top of us. We finally got to the driver to check him out.

I was truly shocked when I opened the door. This was a first for me. The driver was completely naked except for a pair of work boots. There were no clothes to be found inside the truck. I radioed outside for the squad crew to check the bed of the truck to see if there were any clothes in there. I had to repeat the message and then I explained that the driver was completely naked. There was a long pause and then I heard, "Uh.......10-4, I guess."

I then checked the driver and found a pulse and that his breathing was good. There was no blood and he looked normal, except for his nudity. He did, however, have nice work boots. I checked his pupils and they were equal and reactive to light. I then started trying to figure out if he was hurt or asleep. I went around to the other side of the truck and got in on that side to take his blood pressure. That's when I saw that there was half a bottle of whiskey on the passenger side floor. Next to the bottle were about five pornographic magazines. The bottle didn't surprise me a bit but I was a bit surprised by his choice of reading material. I've seen people driving down the road reading a newspaper or map, but I had never seen anyone reading porno and driving at the same time. Then the squad captain, Eddie Shoaf, called me on the radio and told me that the bed was completely empty.

I got in and checked his responsiveness. I yelled at him and received no response. Then I took my knuckles and rubbed his sternum hard. Bullet was standing on the driver's side and saw that the driver woke up when I did that. He said, "He's awake now."

I said, "It's a good thing because next I was going to thump his nuts."

Bullet just laughed. The driver wanted to know what I was doing in his truck. I told him that he had wrecked his truck and I was there to check him out. He didn't believe me until I pointed to the altar of the church. Then I asked, "Where are your clothes? Why are you driving this truck naked?"

"I'm up here from Franklin and working my ass off. Then I get back to the hotel and there was a message from my wife that she wanted a divorce. I went to the liquor store and bought a bottle and decided to get drunk. I went next door to the adult bookstore and bought some magazines and decided to go somewhere and look at them. I guess I must have got a little carried away and took my clothes off."

"Where did you put them?"

"I threw them in the bed of the truck."

"Well, they're gone now. Either you missed or they blew out. Do you have anything else to wear?"

"I don't have anything in this truck. I need to go back to my hotel room."

A police officer was standing behind me listening. He said, "You're not going anywhere, man. You're going straight to jail."

"But what about my clothes? I can't get out of this truck in a church naked. What will people think?"

The police officer asked me, "Can he wear your turnout gear?"

"Hell no! If you think I'm gonna let him free ball in my bunker pants you're crazy. Don't you guys carry anything he could wear?"

"No. Can't we get him anything?"

We finally decided to give him one of our Tyvec hazmat suits to wear. They are disposable and no one will have to wear them again. We figured we would send him a bill for the cost of the suit. After he got dressed, we pulled him out and walked him to the police car. He wanted to take his bottle of liquor but the police officers wouldn't let him. They also wouldn't let him take his reading material, either. I looked at him and said, "Hey, don't sweat it. Just get them to put you in the right cell and I'm sure you'll find somebody to replace your wife." I don't think he found the humor in that because he was cussing me as he was put in the patrol car.

We waited around for the tow truck to come and pull the truck out of the church. Then we shored the wall up a little more and covered the hole with plastic. After that, there wasn't much for us to do so we went back into service. I hated to leave the minister like that but there was nothing else we could do for him. I told him the only thing to do would be to call his insurance company and report the accident. I also told him to get a copy of the police report and he would be able to get the insurance information on the driver. I wished him luck and headed back to the house.

When we got back, we found out that Bullet's favorite team got crushed in the game. We had to ride him a little before we turned in. As I lay down, I asked God to let me get a full night's sleep. I don't know what I did wrong that day, but an hour and a half later, we were toned out to go to a vehicle accident, vehicle in the lake.

Our station is a mile from the lake and we arrived to find three people standing on shore and two police officers in the water around the car. They were trying to look inside but the couldn't see anything because they had no goggles. We walked out on a dock and we were standing about 10 feet from the car when an officer swam over and asked us if we had some goggles. We didn't, but we did have air-packs. We handed one down to him in the water and he was able to put it on. He dove down and was able to see that there was no one in the car. He came back up and he and his partner got out of the water. We called communications and asked for the county dive team to come out and help us get the car out of the water.

I walked back to the parking lot and ran into an off duty firefighter, Scott Randall, who lived across the street from the lake. Scott came over to see what was going on. After we told him, he went back to his house and made coffee for all of us. He had to work the next day but he said he didn't mind. He couldn't let us stand out there all night and not have any coffee. I thought that was pretty damned cool. Most people would have gone back to bed, but not Scott.

Once the dive team and a tow truck arrived, they prepared to dive in and hook the car up to a cable. When they got to the car, they dove down and put the cable on the frame. While they were down there, they received a big surprise. This car that they were preparing to pull out of the lake had landed on top of another car. The second car appeared to have been down there for a long time. The diver radioed the license plate numbers off of both cars back to shore. The first car, a 1999 Acura Integra, was stolen that night from the owner's home. Apparently, the thieves were scared after stealing the car and drove it in the lake. The second car, a 1977 Lincoln Continental, was stolen in 1978. The Lincoln had been down there for over 20 years and no one had ever noticed. You would think with all of the new gadgets on the market for fishing that someone would have found that car by now.

Now we had two cars to pull out. It took half an hour to get the Acura out of the lake. Once we pulled it away from shore, we checked the inside of the car again to make sure there was no one in there. Then I took a Haligan tool and used the point to punch the lock out of the trunk lid. Then I used the flat side to pry the lid open. We looked for stolen items or bodies in the trunk and found nothing. Then we went back to work to get the second car out of the water.

It took us longer to get the Lincoln out. Over the years, it had sunk into the mud at the bottom of the lake and the divers had to dig it out. The divers had to dig for half an hour before they could even get to the rear axle. Then they came out and took the line from the tow truck back to the car. After pulling the Lincoln out, we checked the inside and I popped the trunk on it to see if anything was in there. The rear window was out of the car and when we opened the door we found a seven or eight pound bass in the back seat. Just as a coincidence, the very next day the fire department was going to have a bass-fishing tournament. I just hate that we had to throw it back. I believe I could have won with that one.

The cars aren't the only things we have pulled out of that lake. We have pulled motorcycles, boats, parts of drag boats, and bicycles out of there. We have also pulled bodies out of there, and that's something I hate to do. We have to drag the lake and you never know what you'll come up with. We were out there one day trying to find a body. We found one. The only problem is that the body we found wasn't the one we were looking for. According to the coroner, the first body we found had been down there a couple of days. We eventually found the other body and delivered it to shore.

We finally got back to the station around 4:30 that morning and I decided to take a nap before getting up at 7:00. I was planning on fishing the next day and I didn't want to be completely dead tomorrow. After getting a little sleep, I took a shower and headed to the house after shift change to get the boat. I had some breakfast and picked up Bullet on the way to the lake. Bullet's my partner in these tournaments and I wouldn't have it any other way. He knows that lake like the back of his hand. He's forgotten more about that lake than I will ever know.

We headed out at 10:00 and hit Bullet's favorite fishing spot ten minutes later. He always waits until everybody else leaves before he leaves. He doesn't want anyone to find his spot so we wait and take off like a bat out of hell. I've got a pretty good motor on that boat and we can usually get by people before they can follow us. The only problem is that we have won the last four tournaments and everybody knows that Bullet's the best fisherman on the department. When we went under the bridge, there were nine boats waiting on us. They tried to follow us but I gunned that motor as high as it would go and eventually we pulled away from all of them. They searched for us for about half an hour but couldn't find us. The cove we always go to has a pretty good hiding place and they couldn't see us. Eventually they would have to start fishing because we only had a three-hour time limit.

After we had caught our limit and took what we thought would be a winner we headed back to shore. As we were heading back we passed several teams trying to get some last minute fishing in. We waved to each team and every damned one of them shot us the finger. I couldn't believe it. We try to have a fair competition and they treat us like that. But, we'll get over it. Especially if we win.

We weighed our fish in and the biggest was nine and a half pounds. Definitely a nice catch. We waited as each team came and weighed in. None of the other teams had a catch nearly as big as ours so we took another tournament. We took our prize of $150.00 each and got ready to go home. Of course, we heard a lot of bitching and moaning from the other teams, but they were just pissed because they didn't win. All we could say was better luck next time.

I got home and told Teri about winning the tournament and all that had happened the night before. She was happy that I won but I don't think she liked the fact that I went to the tournament after getting only two and half hours of sleep. I made it up to her, though, I went to bed for a few hours. But, oddly enough, she was even madder after I took a nap. I did make her a little happier later, though. I went out and mowed the yard.

I came in and we all ate dinner while watching a couple of movies she had rented. They were pretty good, and I hung in there for the entire first movie. But I fell asleep during the second one. I finally got up and went to bed. I guess the lack of sleep and all the work I had done the past couple of days finally caught up with me. The next day was Sunday and I figured I could catch up on what I missed after church.

Sunday was a laid back kind of day. We went to church and lunch and then I spent most of the afternoon on the couch watching the race. I took it easy and that evening Bullet and his girlfriend came over. We grilled out and had a nice dinner. Bullet's girlfriend, Jennifer, is the perfect match for him. She's been known to have a short temper herself. We found out that night that she had also used guns in her past. Bullet shot his truck, she shot a car. But, she didn't shoot her own car, she shot an ex-boyfriend's car. He was harassing her and she took a shot at him. She didn't hit him but she did blow the radiator out of his car. Of course, he couldn't tell the police about it because he was in violation of a restraining order at the time. All in all, I think if they had traded places, she would have shot the truck and he would have took a shot at her ex-boyfriend.

The only problem is that they announced that they were planning on moving in together. I'll just have to remember that they are not afraid to use a gun. I would hate to show up at their house and walk into a shoot out. I wonder where I can get a metal detector for the front door of our house. I guess I'll have to get one of those signs that say concealed weapons not allowed in house.

Chapter Eight

Mondays suck. There's no other way to put it. We work twenty-four hours on and get forty-eight hours off so you would think that Mondays would be just another day for us but there is still a difference between Mondays and the other days of the week.

When we got to work, we found out that we had a flat tire on the truck. Bullet got chewed out for it but it wasn't his fault. He had just come on duty and I guess the chiefs thought that somebody needed to get a good ass chewing for the flat tire. The problem is, they had the wrong man. I ended up getting into an argument with the deputy chief because he was blaming Bullet instead of the man that was responsible. All in all, it was a hell of a way to start a day.

We got back to the station in time to get a medical call. It was one of our frequent fliers and I hated going to this house. The elderly couple that lived there were really nice people and I didn't have a problem with them. The problem is that the man has cancer and has what is called a DNR order. A DNR order, or Do Not Resuscitate order, means that if the person is dying or has just died, we are not allowed to do anything to try to resuscitate or prolong their life. It is a tough decision that people have to go through and I can respect that decision. If I had a terminal illness, I would not want to prolong the suffering, either. The problem is that we are trained to save lives. We are trained to do whatever it takes to save a person's life. With that in mind, it is tough to stand there watching a person die and do nothing.

We got on the scene and found the man's wife standing outside crying. I asked her what was happening and she told me that she thought he was dead. I walked in with Rick Johnson to check him out while Bullet stayed outside with the man's wife. He was comforting her while we tried to take vitals. He wasn't breathing and he had no pulse. I checked, then Rick checked and the man showed no signs of life. I radioed communications and told them we had a confirmed code 44, which means we had a confirmed death and I requested a police officer. After I talked to communications, the paramedics showed up and confirmed the death for themselves.

We went back outside and the paramedics told the lady that her husband had died and they stayed with her until the police came. We went back into service and headed back to the station. I hate riding these kinds of calls. It's bad enough that someone has died, but to sit there and do nothing to save them when that is what you are trained to do is even worse. I know these people don't want to prolong their agony but it doesn't make it any easier for us to stand there and do nothing.

We got back to the station and had lunch. None of us said much. The other guys had already had their lunch and were working around the station so we had the kitchen to ourselves. I just wish we could have done more but, like I said, our hands are tied. If we had saved the man we would have gone against the law. I just feel that because we couldn't do anything for him we went against nature.

We didn't have much to do that afternoon. We had a class on streets and found out that Rick was learning his streets pretty well and that the other rookie, Steve Williams, was having a little tougher time learning his streets. Either that or he wasn't riding his streets like he was supposed to. Bullet had a lot of fun ribbing him about not knowing them but, then again, until Bullet met Jennifer he had no life and could ride streets a lot more than most people. We worked on streets for a couple of hours until the tones went off again.

The call came in as an alarm in a dust collector at a local furniture plant. I don't like these calls. You have to get all of the sawdust out of the collector before you can find out if there is a fire. Dust collector fires are dangerous, as well. Sawdust can burn with explosive results.

When we arrived on the scene I had the ladder truck ladder the dust collector. They took a section of hose and a nozzle to the top and hooked into the top of the ladder. They started spraying into the top of the collector, wetting down the top level of sawdust. I made certain they had their ladder belts on and were tied off before we did anything on the ground.

I took my crew and engine 21's crew and we put on full turnout gear and our air-packs. We opened up the bottom of the collector so we could get to the sawdust. The sawdust usually gets packed in there so tight that it will not come out on it's own. Dust collectors usually have a vibrating mechanism built in to shake the collector and allow the sawdust to come out on it's own. The problem is that the vibration devices are electric and I didn't want to increase the danger of explosion. Instead, I had three men beat on the sides of the dust collector with sledgehammers while another man and I used rakes to pull the sawdust away from the bottom of the collector. Bullet was at the truck pumping and Rick was on the hose line in case anything happened.

We worked on that thing for over an hour. I had everybody switch jobs at ten-minute intervals so that we didn't wear out completely. It's hard work trying to get that sawdust out by hand. I also called for the air truck so that we would have enough air bottles and wouldn't run out.

After an hour, we were only about halfway through the sawdust inside the collector. I got on the radio and had fire communications dispatch two more pumpers, the rehab truck and a county EMS unit to stand by in case one of us went down because of the heat.

It was hot as hell out there. The men wanted to take their coats and air-packs off because of the heat but I wouldn't let them due to the risk of explosion. I heard the other trucks check en route as we switched positions again. I went back to raking under the collector.

The sawdust started coming out quicker and eventually started pouring out. The vibration from the sledgehammers along with the weight of the water on top was beginning to work too well. I got on the radio and told the ladder crew to shut their line down but it was too late. All of a sudden, it was like every bit of dry saw dust came out at once. In the middle of that dry sawdust was a spark.

All of that saw dust lit off at once and exploded into a ball of fire. Standing under that collector, it felt like I was hit with a thousand punches all at once. My entire body felt like it was going to cave in. In the force of the explosion, I couldn't run or walk. In fact, I couldn't even fall down. With the force of that explosion pushing in on my body for that split second, I couldn't move at all. Then as quick as that saw dust lit off, the fire went out.

We all just stood there, stunned by what had happened. As I regained my senses, I looked around and saw the other man still standing beside me under the dust collector. The three men who were using the sledgehammers on the sides of the collector had been knocked backwards about ten feet and were slowly getting up. Rick was crawling back to the nozzle and groaning. Apparently, he had the wind knocked out of him. All in all, I thought we were damned lucky. No one appeared to be hurt very badly and, oddly enough, our turnout gear wasn't even burned. Good thing we were wearing our turnout gear. I looked up at the top of the dust collector and saw that the ladder crew appeared to be okay. One man was trying to climb back on the ladder because the explosion had knocked him off the ladder, but because of the ladder belt he did not fall to the ground. The trucks were not damaged, either.

Through the ringing in my ears, I could hear Bullet calling me on the radio. Apparently, he had been calling me for a while because he was saying, "Dalton, you okay?" instead of using correct radio protocol and calling me Engine 18. I could also hear Engine 15 calling on the radio, stating that they could see a ball of flame from Elm Street, three and a half miles away.

I radioed that we appeared to be fine. Then I told the ladder crew to hose down the inside of the dust collector while I took the nozzle from Rick and started wetting down the sawdust all around us. I had the other guys move over to our truck and wait for EMS.

The next two pumpers arrived and I turned command over to Engine 15. They took over wetting down the sawdust. I went to our truck and saw that EMS had arrived and was checking the other guys to make sure they were okay. They checked our vital signs and cleaned some scrapes. No one was seriously hurt and by this time the ringing in our ears had subsided. Most of us were stiff and sore but that was it.

Chief Jones arrived and apologized for not getting there sooner. He had been in a big meeting with the county trying to better our mutual aid policy. I told him not to worry about it. That collector probably would have exploded whether he was there or not. Still, he felt bad about it.

He said, "I know, but I would have liked to have seen it."

I looked at him for a moment and said, "Well, thanks a lot!"

"Nothing personal. But I would have liked to have seen the look on your face after it happened."

"Well, Chief, I can't remember the look on my face but I can say that the pucker factor definitely went up." He gave me a confused look. "In other words, you wouldn't have been able to pull a needle out of my ass with a tractor."

He laughed and told me to go on back to the house. We broke our lines from the truck and let Engine 15 hook into them. We headed back and reloaded the truck. The squad crew came out to the engine room to help us. We then refilled our air bottles and I reluctantly put us back into service. I was just hoping that it would be a quiet night. I told Bullet that I was going to go to bed and told him to wake me when they brought our off-loaded hose back.

After trying to call Teri and getting no answer, I took a quick shower and went to bed. I hadn't eaten anything since lunch but I was too tired to eat. I laid down and went out pretty quick. The next thing I knew, the 7:00 bell went off to wake us up. I went out and asked Bullet about the hose that Engine 15 had.

"They brought it to us around 11:30 last night," he said. "The squad crew and I cleaned it and put it up to dry."

I asked, "Why didn't you wake me up?"

"Hell, Cap, you and the other guys had just been through hell and you were beat. I didn't want to wake you for something as small as cleaning hose. Besides, I had help. Don't worry about it."

I thanked him and went back to my bedroom to get ready to go home. Surprisingly, I wasn't as sore as I thought I would be. I last night's guess that hot shower helped.

After changing clothes, I went to the kitchen for some coffee. I ran into Bill Waddell, the ladder captain. He asked me how I felt. I said, "Like I've been rode hard and put away wet. Actually, I'm a little sore but nothing major. By the way, what happened up on top when that thing exploded?"

He said, "We had a little fire through the access panel on top and a ball of flame rolled up beside us. That was nothing though. I'm just glad we had our ladder belts on or Mike and I would have gone off the top of that thing."

Mike Berrier is a big man. Most doors don't take up as much room as Mike. I looked at him sitting at the table eating his usual six bacon, egg and cheese biscuit breakfast and said, "That would have made quite an impression... on the ground."

He looked up at me with a mouth full of food and said, "Fuck you!" spraying food all over Rick sitting across from him.

Rick said, "Damn, son! Watch it!"

Mike swallowed and said, "Sorry, man. But he's being an ass hole."

"Well what do you call somebody spitting food all over you?"

"Man, I said I'm sorry. I was trying to spread the joy of this wonderful breakfast to you, man."

I looked at his food and said, "You're gonna spread, all right. Spread your fat ass right out of that seat. What the hell do you call that breakfast, anyway?"

He stood up and said, "I'll have you know that this breakfast is healthy. It's healthy because it's all natural. The eggs, the biscuits, the bacon and the cheese. All natural. Just like me."

Bullet walked in for coffee and couldn't get around Mike. He put his hand on Mike's shoulder and said, "Well how about moving your naturally fat ass out of the way so I can get some coffee."

Mike didn't say a word. He just grabbed up what was left of his breakfast and headed out to eat in the engine room. As he was walking through the kitchen door, he hit his hip on the doorframe and stumbled on through. As he walked away, I shouted, "Need us to widen the door for you naturally fat ass?" He just kept going.

We sat around the kitchen for a while, waiting for shift change. As the other shift came in we told them what we had the day before and what we did to get the truck back in service after the calls. They told me they saw the explosion on the news. They said that the news had a perfect shot of the explosion. Ben Jarvis, a captain on the other shift, said he called in to make sure no one was hurt. Then he called the families of all of us involved in the explosion to let them know we were all right.

It's times like this that make the fire department more like a family. I didn't get a chance to try to call Teri again before falling asleep. I felt bad about that but I was out before I realized I hadn't tried her again. But, thanks to Ben, at least she knew I was okay.

After we swapped shifts, I went ahead and called Teri. I told her about the explosion and made sure I apologized for not calling her. She said it was okay and told me to be sure to thank Ben for calling her and letting her know that I was okay. I told her I already did and that I would be home early. I didn't feel like going to my part-time job.

I called them after getting off the phone with Teri and told them that I would be taking the day off. I told Rick Jackson, my part-time boss, what happened the day before and that I would be taking the day off. He told me that it was okay and that if I needed tomorrow off it wouldn't be a problem. I thanked him and told him to tell my supervisor, Dick, to fuck off if he didn't like it. I then went home to take a relaxing day off.

When I got home, no one was there. The kids were spending the day at the YMCA and Teri was at work. I went into our bathroom to change clothes and found out that Teri had been on the Internet. She went to every cartoon site that she could and printed off pictures of cartoon characters that had been blown up. All were standing there blackened by the explosion and looking straight at you with a stunned look on their faces. I just sat down and laughed like hell. I went and called her. I told her I loved the pictures.

She said, "I just thought you might want to know what you looked like when it happened."

"I don't know if that's how I looked but it sure as hell is the way I felt."

"How are you feeling?"

"Not bad. I'm a little sore but nothing serious."

"Well lay down and take it easy today."

"That's probably all I'll do today. I called Rick and told him I wouldn't be in today."

"Was he pissed?"

"Not at all. In fact, he told me to take tomorrow off if I needed it. But I'm going to let you off here. You have a good day, baby."

"You too, Darlin'."

"I love you."

"I love you, too."

I hung up the phone and changed clothes. I then lay down on the bed with the dogs and took a nap. That's about the hardest thing I did both days. I lay around the house and didn't do much of anything but watch television or play on the computer. I picked up the kids both days and made sure we had food to eat but not much else. It felt great but I was starting to get bored. But I knew that wouldn't last long because of having to go back to work on Thursday. I'll just have to make sure not to stand under anymore dust collectors again.

Chapter Nine

I got back to work on Thursday with no problems. I wasn't sore any more and I was eager to do something after the two days of rest. I would rather have been doing something with the family but I guess going in to the fire department is better than doing nothing.

We did station maintenance during the morning and made it through lunch without riding anything. But as soon as we were through with lunch, the tones went off and we were dispatched to a car fire. When we arrived on the scene the fire was almost completely out. There was a little smoldering in the carpet of the trunk and the CD player that was mounted back there but not much else.

The first thing we did was disconnect the battery and I pushed the hot line down beside the battery and the negative line behind the battery. Then we pulled the carpet out and took the CD player out along with the back dash and one speaker that was a little melted. Other than a little bit of scorching around the rear side light of the car and a lot of dry powder from an extinguisher, you could not tell that anything was wrong with the car.

To be on the safe side we checked everything on that car. We pulled up the carpet and checked out the wiring harness under the carpet beneath the back seat and pulled out the molding around the rear window behind the back door. All four doors of the car were open and the trunk lid was up. The car had fold down seats in the back and both were down. The seats had an upholstered frame around them that would not fold down. We checked all of the wiring up to the frame around the seats and throughout the car that we could check without totally destroying the car. We also felt for heat in all parts of the trunk and the rear seats every five minutes.

The owner told me that the night before he had run into something and put a small dent on the side of the car and had broken the rear sidelight. He said he was trying to change the light when he heard a loud pop. At the time of the fire he was listening to the CD player while he was working. He told me he didn't know whether he had pushed the light too far into the compartment and shorted something out or if there was a loose wire. All he knew was that he pushed the screwdriver in, heard a pop and the saw the smoke coming from the trunk. He saw that the CD player was on fire and reached in and ejected the six CD cartridges to get the CDs out. His neighbor came over with a fire extinguisher and they put it out.

I called the on call fire inspector to see if he wanted to come up and take a look at the car. He told me that it would not be necessary for him to come see the car if the battery was disconnected and if I felt that the fire was completely out. After a total of thirty-five minutes, I told the owner to keep an eye on the car and call us back if there was any problem. Before we left, the man asked if he would be able to drive the car. I told him that he would need to have the car towed. If he was to hook the battery back up, he could possibly start another fire.

We went back to the house and started a class on our foam system. We were there for thirty-five minutes when we were toned out again, this time for a full alarm to the same address as the car fire. We arrived on the scene to find the car fully involved and impinging on the house. As I pulled off an attack line I had Engine 15 catch a hydrant a block away. Rick and I pulled one line and worked our way to the rear of the car. I hit the rear of the car to knock the flames down and then turned and hit the siding on the front of the house. As far as anyone could tell, the only thing wrong with the house was a broken window directly above the car, the vinyl siding was melted and the vinyl garage door was melting.

We got the fire in the trunk and passenger compartment out in a matter of minutes and started trying to pop the hood. The handle was gone on the inside of the car, of course, so we had to use a Haligan tool and pry bar to get the hood up enough to get a set of bolt cutters in there to cut the latch. When we got the hood up, we noticed that the only fire in the engine compartment was located right at the rear driver's side of the engine, where all of the wiring goes through the firewall. We knocked that fire out and started assessing the damage.

The first thing I looked at was the battery. I remembered what position I had pushed the battery cables when I disconnected the battery the first time we rode the car. The cables were in the same position as we left them. We then went to work to try and find the cause of the fire. The on-call fire inspector, John Robertson, arrived and we started checking every possible lead that we could think of. We checked all of the fuses that were in a second fuse panel under the hood. We found that the fuse for the lights on the car and the fuse for the CD player were both blown out. All other fuses were okay. We then started following the wires from the back of the car to see if there was a problem with how the wires were running through the car. We could find nothing wrong with the wiring system and we didn't know if there was a capacitor that could have fired off a charge through the wires that were burned earlier.

We found out that the only damage to the house was one broken window, one melted vinyl garage door and the vinyl siding was melted off the garage section of the house. Inside, the only damage was a slight smell of smoke. Another crew checked the attic of the house and there was no damage there.

I couldn't figure out why our twenty-four foot extension ladder was on the front of the house. I found out that after arriving on the scene and not seeing a trunk lid,

Chief Jones had a crew ladder the house and look for the trunk lid on top of the house because everyone was talking about the car exploding. But before the crew started up the ladder he realized that the trunk lid was made out of aluminum and had melted.

After we thoroughly checked the car out, we went back to the station and loaded our truck back and cleaned up the hose we used on the call. Then Chief Jones and John Robertson came to the station and we talked about the earlier call and the last call on that car. We could not figure out how the second fire started. The owner said that he was sitting there and noticed a few wisps of smoke coming from the car. He went inside to call the fire department and when he came back out the car just took off in a matter of seconds.

We were trying to figure out the cause of the second fire. John found out the man had a degree in electronics and had to be very good at it because his job was repairing the electronic truck scales along the side of interstate highways. We then tried to find something that he could have done to start the fire.

We knew that he did not hook the battery back up because the cables were just where we had left them. If he had hooked them up, he disconnected them again and put them where we had left them. Inspector Robertson had picked up a wiring diagram from a local car dealership so that he could see if there was a capacitor in that line. Away from the engine compartment, there were no capacitors in the wiring system. But none of us could figure out how a fire could rekindle in a car seventy minutes later without help.

I don't know if we will ever find the exact cause of the second fire, but chances are there will be a lawsuit over it, either by the owner or his insurance company. People are so eager to sue that they don't think of the fact that we saved other things. In this case, the house only had minor damage because we were able to quickly get the fire out. That probably will not matter to the owners. Had we not been there, the house would have been damaged a lot worse. We can pull people out of a house fire and save their life, but if they break a fingernail when we pull them out, they'll sue us. It never fails.

We did ventilation on a house fire once. We cut a four-foot by four-foot hole in the roof and reached through with a pike pole and knocked the same sized hole in the ceiling. We do this because it eliminates a lot of the heat and smoke and because it eliminates the danger of a backdraft, or smoke explosion. Another thing the ventilation hole does is it has a tendency to centralize the fire and keep it from spreading as fast because of oxygen. The main source of the oxygen for the fire is from the ventilation hole so the fire will stay close to that hole to stay alive.

The owner of the house on this fire saw us cut the hole in the roof. When he saw us take a chain saw to the roof of his house, he went ballistic. Once the fire was out and he found out that only two rooms and the roof on that end of the house needed to be repaired, he was still pissed. We received court papers and were sued for causing unnecessary damage to his house. Of course, the case was thrown out before it came to trial. The judge called the owner a dumb ass and told him that he should be thankful that we were there and were experienced enough to keep damage to a minimum. Of course, it didn't hurt that the judge was a volunteer firefighter and knew what he was talking about.

After we got cleaned up we decided to go to the grocery store and get food for dinner. After doing our shopping we were heading back to our truck when a car pulled up. The driver, a woman in her early twenties, rolled down the window and asked if we could help her out. I gave the bags to Bullet and walked over to her car and asked what I could do to help her. She said, "I have a problem with ben wa balls."

Ben wa balls are sexual toys that are inserted into a woman's vagina and the vibrations are supposed to give the woman pleasure. I couldn't see how anyone would have a problem with them, so I had to ask, "What kind of problem are you having?"

With that she pulled her dress up and said, "I have one stuck in here. Can you get it out?"

I wasn't sure what she expected us to do. I knew I wasn't going to try to get them out on my own. Besides, this was too good to keep to ourselves. I called fire communications and had them send a county EMS unit, non-emergency traffic, to our location. I asked the woman if she was in any pain.

"No I'm not in pain. I got one out but I can't find the second ball. I don't know what to do."

"Well, I called for an ambulance and they may know something to do."

"You called for an ambulance? I can't believe this. All I want is for somebody to pull this damned thing out of me and you go calling for an ambulance. This is so embarrassing. I can't believe you would do that."

"Well, ma'am, you came up to me, a complete stranger, to see if I could get that ball out of you. I don't see what two more strangers will mean."

"I guess you're right. Just out of curiosity, are the guys on the ambulance cute?"

"Oh, they're adorable." That's the point that I knew that I wouldn't be touching her. I looked over at Bullet and said that we wouldn't be touching her. He gave me a disappointed look but then I whispered, "She could have something Ajax won't wash off."

He thought for a moment and said, "You're right."

When EMS showed up, I told Jack Walters, the head paramedic, what we had. I gave him the information that we had. He gave me a dirty look and said, "And you guys just had to call us, didn't you?"

"Well, you guys are the medical experts, not us." I laughed as we walked over to the car.

Jack talked to the woman for a few minutes and then got his gloves out. He squirted a lubricant on his gloved hand and asked us to step back so he could examine the patient. We all stepped back, but not so far as to be out of earshot. As Jack was examining her, we noticed that her breathing started becoming erratic. She was panting and started moaning and finally Jack pulled his hand out and stood up. She looked up at him and he told her, "Ma'am, you're just going to have to go to the hospital. I can't get this thing out and you're starting to enjoy this way too much."

She was disappointed but walked over to the ambulance and got in the back. Jack asked me if I wanted to ride in with him. I said, "No I think we'll give you two a little privacy."

"Oh, kiss my ass!"

"Go get 'em, big boy." With that, we were all blowing kisses and making obscene noises as he climbed in the ambulance. After the doors were closed, we could see him shooting us the finger as they pulled away. We all laughed about it for a few minutes and then decided to head back to the station.

We fixed supper and everyone ate all they could eat. We sat around talking about the woman in the parking lot and everybody was laughing like hell. Mark Walters is Jack's brother and the driver of the ladder truck. He couldn't resist it. He got on the phone and called the EMS base and asked to speak to his brother. He was on the speakerphone and when Jack answered the phone, he said, "Hey there, Stud, I hear you got lucky today."

He yelled, "You tell Dalton I'm gonna kick his sorry ass. All the way to the hospital she kept begging me to put my hand in there again and work on getting that ball out."

I asked him, "Well did you ever get it out?"

"Yeah, the ER doctor gave her a shot of muscle relaxer and it came out a couple of minutes later."

Bullet yelled, "He gave her a shot in her twat?!?!"

"Yeah. I don't think she enjoyed that too much and I don't think we'll be seeing her any time soon."

I used my most sympathetic voice and said, "Well, Jack, you'll just have to get back out there. Don't worry about this breakup, you'll find another girl."

With that, we heard him say, "Fuck you!" as he hung up the phone. We all fell out with laughter and we made more jokes about it. Bullet was still standing there with a shocked look on his face. "I can't believe they gave her a shot in her twat."

I said, "Well, damn, Bullet, if you two would have been alone I bet you would have gave her one, too."

"Yeah, but mine don't hurt near as much."

The running joke about Bullet is that his manhood is about the same size as a small caliber bullet. Without hesitation I said, "Well, being hung like a pimple, that's understandable."

He grabbed a biscuit off his plate and threw it at me. Everybody was calling him Little Bullet and picking on him about his "Pop Gun." Frank Jackson started picking on him about the size difference between black men and white men. Frank is a black firefighter and is always going around talking about that. Bullet looked at him and said, "Frank, just remember, I've seen you in the shower, too. And that myth about black men being bigger than white men? You've proved that myth is a lie."

That took the heat off of Bullet and put it on Frank. Frank finally said that he wasn't going to put up with this shit and started to walk out of the room. That's when I reminded him that it was his turn to do the dishes. He reluctantly walked back into the kitchen. He looked at me and said, "I know that you ain't the biggest man in the world, yourself."

I said, "I have enough to make my wife feel good and I can lick my eyebrows. According to her, I don't need anything else."

He looked over at Bullet and said, "I just want to know what the hell you're looking at me in the shower for?"

Bullet said, "Well, when there wasn't much there, I had to make sure I wasn't in the shower with a woman." We all walked out of the room laughing and calling Frank sweet cheeks and sugar. I know one thing is for sure; Frank doesn't take a joke as good as Bullet does. I figured we would give him a little time to pout about it then everything would be okay.

We went into the classroom and watched the news. A couple of guys were studying for promotional tests and others were kicking back and relaxing. We have two men on our shift that are taking the paramedic courses at the local community college and were studying for that. We were able to kick back and take it easy for a while. It's always good when you can relax on this job, and when that time comes, you take advantage of every second. Of course it never lasts very long. The tones went off and sent Engine 18 and Squad 18 to a two vehicle accident with possible pin-in. They also told us that gasoline was pouring out of the back of the vehicle. We were en route to the scene fire communications came back on the radio and said that witnesses are claiming the car is now on fire with the victim still pinned in. At that point, we could see the smoke over the treetops and knew that we would have to hustle. When we got on the scene I pulled an attack line while Rick and my other firefighter, John Williams, put their air-packs on. I could hear the victim screaming when I pulled the line off the truck. Bullet helped me clear the bed. When the hose was off and I was headed toward the cars, he charged the line. Before the first drop of water hit either car, the screaming had stopped. The driver, still pinned in the front car, was dead.

There is no sound worse that hearing someone screaming while being burned alive. When you hear it, you want to make everything all right but you know that you can't. Even if we had arrived a minute sooner, there is probably nothing we could have done to stop the driver from being killed. Car fires never burn slowly. It takes a few minutes for it to build, but when there is gasoline leaking from one or more cars, the time it takes for the fire to destroy everything in its path is minimal. When you hear that kind of scream, wish you could be deaf for a while. You wish you couldn't hear the screaming but it is there. It echoes through your head long after the call is over. You close your eyes at night and all you can see is the car fire and all you can hear is the screaming. You open your eyes and see the darkness of your bedroom. When you close them again, it all comes back to you once again.

Once the fire was completely out, Chief Jones arrived and asked what had happened. I gave him a report of what I knew at the time. Witnesses told me that the victim's car was headed down the hill of the street directly behind both cars. He had stopped for the stop sign at the T-intersection. The driver in the rear car came speeding down the hill behind the first car and slammed the first car from behind, driving it through the intersection and into the guardrail across the street. The driver in the rear car was drunk and walked away. The driver in the front car had his foot pinned between the clutch pedal and the sidewall of the car and could not get out. When the cars started burning, one man tried to pull the victim through the window but could not free him in time and had to back away before he was burned.

Chief Jones got on the radio and asked that the CISD team be sent to wait for us when we returned. CISD, or Critical Incident Stress Debriefing, is a big help to firefighters, EMS personnel and police officers around the country. I have dealt with these people before and they are great. When you have a bad call it is good to talk about it. To get your feelings out in the open. You know it's not your fault the people are dead, but if you let these feelings and problems you have to deal with build up inside you, you are in for a world of problems. It can affect your home life, your work, and it has driven some people to drinking and drugs.

Once the police had finished their investigation, I found out that the victim was only sixteen years old. He had only had his license three days. I asked them to keep the drunk driver on the scene. When one officer asked me why I said, "So the drunk motherfucker can see what he did when we pull that victim out of the car. I want him to see the burned body. In fact, it would be nice if he had to zip the body bag. Just so he can see what his fucking stupidity has caused."

The police kept him on the scene and made him sit there and watch everything. We cut the door open on the car and pried the pedal away from the victim's foot. Then we had to peal the victim from his seat and lay him on the medical examiner's gurney. We zipped the body bag up to the chest of the victim and then rolled him past the drunk driver. I stopped in front of the drunk and as I zipped the bag the rest of the way I said, "Do you see, motherfucker? Do you see what you caused? This was a sixteen-year-old kid and now, because of your fucking stupidity, he's dead. All because of you. You are the one to blame here. I just hope that one day you can feel what this kid had to go through. Maybe then, you'll understand."

The drunk couldn't say anything. He just stood there and stared at the victim. I went ahead and zipped the bag all of the way up and rolled the victim to the medical examiner's van. We put him in the van and went back to get our gear together so we could get the hell out of there. I looked over and the drunk was on his hands and knees on the curb puking his guts up. I just sat there looking at him and all I could think was that there is no justice in the world. The victim dies and all he does is throw up.

We headed back to the station to reload the truck and clean the hose we used on the fire. We helped the squad crew get their gear back in order from the extrication. Then we went in and cleaned ourselves up. Chief Jones arrived at the station and radioed fire communications to take Engine 18 and Squad 18 out of service until further notice. We then all met in the kitchen and talked to the CISD team.

Usually, for me at least, all I have to do is talk my feelings out with someone and I am fine. I don't let those feelings bottle up inside me. I can either talk it out with a CISD member or I can talk about it with my wife, Teri. She knows the drill. She has gone through CISD debriefing herself from of being a nurse in the ER. She knows about the bad calls and it seems she knows exactly what to say to make me feel better. I can talk to her about any call I run and she understands because she has been there. Even if she has never seen anything that can compare to a particular call, she will still listen, and that is one of the greatest things she can do to help. Just listening to me. In fact, even if I do talk to a CISD team member, I still usually call her on the phone and talk with her, too. I guess it's just hearing her voice and knowing that she loves me that makes everything better. I just know that without her, I would go nuts.

Chief Jones gave each of us a chance to talk with CISD team members and, after a couple of hours, asked if we were ready to go back into service. We all said we felt better and he put us back into service. All I wanted to do was to talk to Teri and then go to bed. I called her and talked things out with her and, as usual, she knew exactly what to say to make me feel better. I told her I loved her and then went to bed. This time, when I closed my eyes, I didn't hear the screams and I didn't see the fire. Talking it out really does help.

The next day, I went to my part-time job and had to hear a bunch of shit from Dick Randolph, my supervisor. He was pissed because I had taken two straight days off. I just ignored him, which pissed him off even more. Every time he would make a smart-assed comment, I would just ignore him. Dick wanted to pick a fight and I didn't feel like fighting. He was still pissed about the dog shit in his truck and the ticket he received for littering. He had to pay a $250.00 fine and the judge would not let him explain that he didn't know the bag was back there. The more he tried to get at me, the more I ignored him. The more I ignored him, the madder he got. By the end of the day, he was stomping around like a mad bull and I couldn't have been happier for him. That just made my day and I didn't have to do a thing.

I picked the kids up and we went home to have some fun. On the way, we stopped off at the store to buy some school supplies. The first day of school was quickly approaching and they would need school supplies and clothes. Teri and I always had an agreement, she would buy the clothes and I would buy the school supplies. I don't do clothes very well. After fifteen years, I have finally learned Teri's size but I wouldn't know where to start with clothes for the kids. I just let her handle that part.

After buying notebooks and pencils and the like, we stopped off and got some burgers and headed home. We met Teri in the driveway and the kids told her about everything they had bought for school. The odd thing is, they actually like school. I didn't really like it that much when I was a kid. Not until I went to college and started partying and having fun. That's when I didn't want to leave school. But Joey and Nikki like school and look forward to going back. I guess they get bored not having as much to do in the summer. Of course, I keep telling them that they will look back on these summers when they "had nothing to do" and wish they could have them back. They don't believe me, but it will eventually happen.

After dinner, we watched a couple of movies. The next day was Saturday and I wouldn't have to go to the fire department or my part-time job. I would be able to spend the day the best way I know how, with my family. Joey had a baseball game to play in the morning and Nikki had a couple of things she needed to get for a craft project at the YMCA, but other than that we were open for anything.

Joey won his game. He did pretty good in it, too. He had two doubles and a home run. He was pretty happy with it and was on top of the world. He plays second base and was part of a double play that won the game for them. I was the proud Papa.

Then we went shopping for Nikki's craft supplies. We bought her stuff and I found some wood working tools for Joey and me. Teri is into making jewelry and she even found some supplies that she needed. We headed to the house and decided to kick back and take it easy.

Most of the housework around here is done during the week so we don't have to worry with it on the weekends and can relax. I do the yard work during the week and all that involves is mowing the yard. Teri likes to tend to her flowerbeds so I let her handle that part. I don't worry with bagging the grass or anything like that. I fertilize and seed the yard in the spring I don't but that's about the extent of it. I don't worry about the yard too much. Some people would say that I am lazy about the yard but since I don't get a lot of complaints, I don't worry with it. If it ain't broke, don't fix it.

In the afternoon, we took it easy and sat outside watching all of the neighbors working around their houses. I would much rather watch people work than actually have to do the work myself. To me, that's a lot more enjoyable. We sat outside and took in some sun. Joey and I pitched the baseball back and forth for a while. Then I decided to grill out. There is nothing more enjoyable to me than relaxing with my family. As a matter of fact, grilling dinner was the hardest thing I did all day. We finished off the day with a couple of movies and went to bed.

Sunday, we decided to skip church and head to the mountains. We hiked up a trail to one of the peaks and walked all around the pinnacle. No one is allowed to go to the very top of the mountain because it's reserved for the hawks that live up there. There are more than 500 hawks living on top of that mountain. Along the trail, you can look up and see them taking off and floating on the wind currents. It's as if they are moving in slow motion. They float out a little ways and will stop in mid-air. Then they will float back to a tree on top of the mountain. Along the trail, you can see where the hawks have cleaned out their nests and you can find feathers, skeletal remains from their meals, and old pieces of nests.

We stayed on that trail for several hours, looking around and exploring caves. Fortunately, we took water with us because it was pretty warm up there. But the winds took care of keeping us cooler. We went around that pinnacle twice and then headed back to the Jeep and down the mountain. Joey wanted to stop off at the guest center to see if they had any information on hawks. We found a few pamphlets and headed home. It's an hour drive so the kids sat back there reading. Once we got home, we relaxed and took it easy for the rest of the day.

We ordered pizza for dinner and played a couple of games that night. I went and got everything ready to go to work the next day and then cuddled up next to Teri after running the dogs off the bed. All in all, it was a good weekend and I hated that it had to end. But tomorrow I would have to go to work and be away for another day. Such is life.

Chapter Ten

One of the best things about being in the fire service is the camaraderie we feel among other firefighters. What makes it even better is how we can find humor in the most unlikely places. As an example, take a call that B shift had yesterday. Apparently, it was one of those crazy Saturday calls.

The call came in as an attempted suicide. It seems that a man in our first call area had had enough of this world and wanted to end it all. Of course, after attempting suicide he changed his mind and called for an ambulance. When they arrived on the scene, the police were trying to get information from the man about the pills he said he had taken. The man couldn't talk from crying too hard.

Tony Jester, the squad captain, started talking to the man and did his best to calm him down but with no luck. Finally, Tony asked him where the bottle was, thinking he could get some information from the label. The man handed Tony a flat, round case and managed to say, "This is it."

Tony opened the case and saw that there were no pills inside. He asked the man if he had taken all thirty-one pills and the man nodded yes. Tony said, "I don't think you have anything to worry about, sir. I think you'll be just fine."

"You mean, I'm not going to die?"

"Well, I've never heard of birth control pills being fatal. On the upside, you probably won't have to shave for a month."

We all fell out after hearing that story. We probably shouldn't have been laughing at a man who tried to commit suicide. But we were not laughing at his reasons. We were laughing about the fact that he was not bright enough to check the pills out before he took them. Damn near every adult knows what a birth control pill case looks like. It's not like they are very hard to find. Not to mention that the thirty-one numbered pills should have tipped him off as well. I'm just thankful that he never found her diaphragm; otherwise he could have had some serious problems.

After lunch, we were toned out to respond to a boating accident at the lake a mile down the road. Engine 18, Squad 18 and Truck 18 were instructed to take the boat to the marina to respond to a boating accident, boat versus human. I thought to myself, "I wonder who won." We arrived at the marina and were preparing to put the boat in the water when a bystander said, "Here they come." Apparently, when the accident happened, a friend of the victim took another boat to the docks at the marina and called 9-1-1.

The boat pulled up to the dock driven by a woman. On the rear deck of the boat, her husband was laying with towels over both legs. I asked what happened. The woman told us that the man had been water skiing and fell. Instead of making a wide circle to pick him up, she backed the boat up to him and hit him with the prop. The prop cut four deep gashes in his upper thighs, two on each thigh. Half an inch higher and he would have gone from a rooster to a hen.

We checked his vital signs and tried to stabilize him as best we could. He was starting to go into shock and we didn't want that without EMS there. We dressed his wounds and kept him warm. Fortunately, when the prop hit his legs, no major blood vessels were cut. In fact, he barely bled at all, even though the cuts went nearly three inches deep.

EMS arrived and packed him for transport. As he was being moved to the ambulance, he called his wife over. "Honey, it's all right. I'll be okay. Don't worry about me." He kissed her and we started toward the ambulance again when he wanted to talk to his friend waiting by the door of the ambulance. He said, "Don't let her drive my truck. You take her to the hospital. Ain't no telling how bad she could fuck that up. You drive my truck, not her." His friend agreed and we loaded the man in the ambulance. As we were closing the doors, I heard him tell the paramedic, "That's the last time I ever let her drive the boat. She could have killed me."

As the ambulance pulled away, I told the wife not to worry about the boat. I told her we would secure it to the dock and we would give the key to the man in the office. She thanked us and left with the friend driving the truck. We cleaned everything up and secured their boat. After turning the key in to the man in the office, we went back into service and headed back to the house.

Everything stayed quiet the rest of the afternoon, but during rush hour, we were toned out to a vehicle accident involving a tanker truck. As we were pulling out of the station, fire communications called back and said that the caller advised that there were no injuries but gasoline was pouring out of the side of the tanker in a heavy stream. I advised them to send us the Haz-Mat team and to notify the local environmental clean up agency that we needed a clean up team.

When we arrived, we found everyone involved to be uninjured. A pickup truck was changing lanes and pulled right into the side of the tanker. The police department was already diverting traffic in both directions. We started by foaming the entire area around the tanker and pickup truck. The tanker driver told me that he had already called his boss and they were sending an empty tanker to off-load what was still in his truck.

Rick and I went to see how big the hole in the tanker was. When the pickup it, it tore a six to eight inch gash in both the outer shell and the inner shell. We didn't have anything on the truck big enough to plug a hole that size.

Then I looked around and saw that the front right tire on the pickup truck was shredded. I told Rick to go get the duct tape as I looked for a large piece of rubber. When he came back, we jammed a piece of rubber in the hole and duct taped it into place. That didn't stop the leak completely, but it slowed it down considerably.

Once we were finished, I walked around to the other side of the tanker. There is a creek there that runs parallel to the highway. Fortunately, because of the curb at the side of the road, the gasoline had not started towards the creek. We took all of the absorbent we had on the truck and the squad and started to dam the curb in case the level of the gasoline rose too high. That's when Bullet came on the radio and said, "Pumper 18 to Engine 18, 'Hot Shot' Trooper just arrived." I told the guys to keep damming the curb and I went back towards our truck.

"Hot Shot" Trooper is a real ass hole. He hates firefighters with a passion. But it's okay because we hate the sight of him, too. Hell, most of the state troopers hate him more than any firefighter ever dared to. One problem with "Hot Shot" is that he always shows up with a lit cigar hanging out of his mouth. We have warned him, even begged him, to stop smoking around vehicle accidents but to no avail.

This time, I was ready for him. I took the nozzle and pointed it straight at his door. When he stepped out of his cruiser, he was smoking that stogie. I yelled, "No Smoking!!" Then I opened the nozzle and soaked him from head to toe. He was pissed off but I was able to accomplish three things with that short burst of water. First, I extinguished an open flame near a gasoline spill. Second, I made sure that he would think twice about smoking around one of our calls again. And third, I was able to piss him off...big time.

He started towards me when another trooper, who saw the entire incident, cut him off and told him that he had been warned about smoking around wrecks, but would not listen. The trooper, who outranked "Hot Shot," told him to go back to their base and change clothes. After "Hot Shot" left the scene, the second trooper walked over to me laughing his ass off. He said, "You know, that's just what that sorry motherfucker deserves. Maybe next time he'll listen." He walked away still laughing. Nothing was ever said about the incident after I reported it to Chief Jones.

The Haz-Mat team arrived and plugged the hole in the tanker. We all waited around while the remaining fuel was off-loaded. The Haz-Mat team went back to quarters and I sent the squad back, as well. My engine stayed on the scene so that we could assist the clean-up crew and stand-by on the scene while they cleaned the mess up.

Chief Jones came out to the scene and brought us some burgers and drinks. Of course, Bullet had to complain about that. He always bitches about the food being cold, but Chief Jones was ready for him. He reached through the window of his car and pulled out a bag. He told Bullet, "This one's yours."

Bullet smiled and said, "Hey, this is still warm."

Chief Jones waited until Bullet had taken the first bite, then told him, "I hope that's warm enough for you. I sat on it all the way up here." I thought Bullet was going to spit the bite he was chewing on out, but he thought for a second, shrugged his shoulders and then took another bite. Chief Jones looked at me and said, "That man would eat anything."

Bullet said, "Well, Chief, I can't help it. Right now I'm so hungry I could eat the ass end out of a menstruating porcupine."

Chief Jones was pretty well disgusted with that, so I decided to change the subject. I told him what happened when "Hot Shot" Trooper showed up. He got a good laugh out of that and told me, "Don't fucking worry about it. He's been warned about smoking so many times, it's ridiculous. Maybe now he'll know we're serious about it."

I said, "I just hope his supervisor doesn't call complaining about it."

"Well, his supervisor is my brother-in-law. If he calls you and gives you a bunch of shit, you tell him to call me. He fucks with me, I'll divorce his sister and let him put up with the bitch."

We all laughed. But we knew that Chief Jones talked a good game but he was harmless. We all knew that he loved his wife and wouldn't do anything to hurt her. He just wanted to talk tough. He likes to let us think that he is the boss of his house. Of course, we all know different, but we let him live in his fantasy world. Besides, no one else would have him and he wouldn't have anyone to shine his old bald head if they ever split up.

We stayed on the scene until 1:30 the next morning. It took awhile to get that mess up and they were working as fast as they could. I was wishing that they would finish so we could go back to the house and go back to bed. All we did was sit around in case something happened. Nothing did, but we wanted to be there just in case. At 1:30 we loaded up and went back to the house. We reloaded the truck and cleaned up the off-loaded hose. Then we refilled our water tank and our foam tank. Once that was done, I called fire communications and put us back into service. Then I went to bed so that I could get a little sleep before going to my part-time job that morning. As I was going to sleep, I thought to myself, "All I wanted was a quiet Sunday, and this had to happen."

The next morning, I slept through the 7:00 bell. Bullet came to my room at 7:45 and woke me up. I got dressed and headed out for line-up. We told A shift what we had and then we swapped out. I went on to my part-time job and basically screwed around all day. They didn't have anything for me so I swept the floors and organized the toolboxes. There wasn't much to do so I settled for busy work. Good old Dick didn't say a word to me. He was still pissed that I wouldn't fight with him the other day so I was getting the silent treatment. Ordinarily, I don't like the silent treatment, but in his case, silence was truly golden.

I did the usual routine on the way home. I picked up the kids and headed to the house. I wasn't sure what to do for dinner, so I called Teri and we decided to meet in town. We went out for pizza and, afterwards, we went shopping so the kids could get some school clothes. I kept teasing Joey about having to go back to school and how much he would miss these summer days. He took it in stride and said that he was looking forward to going back.

We finished our shopping and headed home. We kicked back for a while and then decided to head off to bed. I was taking the next day off from my part-time job because it was the kids' last free day. The day after tomorrow, they would be starting a new school year and this would be our last chance to spend a weekday together for a while. The kids said they wanted to go fishing so we got all of the gear out. Oddly enough, Nikki is the one who always wants to go fishing. I guess she has a little bit of tomboy in her. Of course, she won't bait her own hook or handle the fish, but other than that, she loves fishing.

After the kids went to bed, I went into the kitchen and Teri and I packed a picnic lunch for tomorrow. Then we went on to bed and I fell asleep pretty quickly. Of course, I didn't go to sleep without doing a little cuddling with her. Then Teri started rubbing my back and that was all she wrote. I was out like a light. When the alarm went off the next morning, I woke to find her laying with her head resting on my back. Even though she needed to get up and get ready for work, I let her sleep for a few minutes longer, enjoying the feeling of being next to her. Then I decided I had better wake her up. As she got ready, I woke the kids up and fed them breakfast. Once we had eaten our breakfast, I had them go get ready to leave while I kissed Teri good-bye. She went to work and I got ready to go fishing.

We left after loading the Jeep and went to a nearby lake. We went out and started fishing under a big tree. It was nice out there. It wasn't too hot, there was a nice breeze and we had fun watching ducks swimming on the lake.

We stayed out there for about six hours and decided to go home after catching our limit. We stopped off for ice cream on the way and pigged out on it in the back yard. I went in and made Mexican food for dinner. When Teri came home, we had a feast fit for a king. Well, I don't know if it would be fit for a king, but we liked it. We did the works: nachos, tacos, and burritos. The owner of Taco Bell would have been proud.

After letting dinner settle, we had the kids start getting their things ready for school tomorrow. They hadn't put their backpacks together yet and they needed to figure out what to wear. Especially Nikki. She is so slow to decide what she wants to wear that sometimes it seems she needs to start picking her outfits two or three days before. Joey doesn't give a shit. Give him something to cover his butt and a shirt with something written on it and he's happy.

Once the kids had everything ready for the next day, it was still early so we went into the kitchen and played cards for a while. Just something to do with the family before bedtime. But, alas, the time came for them to go to bed. They wouldn't be able to stay up late and watch Letterman but I showed Joey how to set the VCR so he could record it.

I know that Letterman may not be the best thing for a ten-year-old boy to watch, but it beats the hell out of Barney. And I know that I would worry more about him watching Barney than I would Letterman. Some of the snooty women at church criticize us for letting him watch it. But I usually cut them short by telling them it's either Letterman or porn. Which would they prefer that he watched?

Teri and I decided to call it an early night, as well. She had had a rough day at work and was tired. After we showered, I rubbed her feet and legs. Then I had her turn over and gave her a back rub until she fell asleep. It was a little more work than cuddling but she enjoyed it and so did I. I wouldn't be able to do this for her tomorrow because I had to work, but at least I could help her out while I was there. After she fell asleep, I decided to do a little reading and then fell asleep, myself.

Chapter Eleven

There were some changes made in the fire department today. We got a new Battalion Chief today. Frank Williams has been on the fire department for ten years and has now made it to the level of Battalion Chief. I don't begrudge anyone for making it to that level in such a short time, but I don't like how Chief Williams did it. Nobody likes a suck-up and in Chief Williams' case, he should have Shop-Vac tattooed on his ass.

That morning, Chief Williams came to the station to talk to us. I wasn't able to attend the meeting with everyone else because I was in a meeting with the Suppression Chief, Bill Richards. We were having a meeting about the new trucks we were ordering and I had all of the specifications that the committee came up with before I went on vacation. After the meeting, I walked into the kitchen where Chief Williams was talking to a few of the men. When I walked in, he came over and said, "Well, Dalton, are we going to have any trouble working together?" I said, "I don't think we'll have any trouble working together. I don't have any respect for you but I don't see us having any problems."

All he could do is stand there with his mouth open. Bullet said, "Well hell, Cap, none of us have any respect for him. But we'll just have to deal with it."

"Well that's what I plan to do. I don't give a shit what happens. I'll do my job as long as he does his and doesn't interfere with mine."

The tones went off and we were saved by the bell. We left Chief Williams standing there with his mouth hanging open. When we got on the truck, Bullet asked, "Well, do you think he's pissed?"

"Doesn't matter. As slow as he is, I'll be retired before he realizes he's been insulted."

We were toned out to a commercial fire alarm at a local hardware store. The first due engine arrived and returned us to quarters before we got there. It turned out to be a false alarm and they didn't need us. On the way back, we stopped for groceries since we were near a store. Then we headed back to quarters to get started on our station clean up.

That afternoon, John Williams came by for a cup of coffee and to talk. John was my first captain when I came out of training. He retired fifteen years ago but comes by the station every once in awhile to catch up on gossip. Today he came by with a great story about his son, Terry. Terry is a captain at Station 8 and is as good as gold. He loves guns and likes to go target shooting and hunting. John told us that he had taken Terry's black powder gun for the weekend to do some work on it. The stock needed to be fixed and it was in dire need of a good cleaning. Once John had finished with the gun, he tested it out in the back yard and found everything to be in working order. He then called Terry and told him he could come and pick his gun up when he was ready to.

That evening, Terry stopped by John's house and took the gun out to his truck. He leaned the gun against the passenger side of the seat with the barrel pointing down. After a few minutes of small talk, Terry decided to take his gun home. He got in his truck and backed out of the driveway. John said, "As he turned to go up the street, I heard a loud boom and saw smoke coming from Terry's truck. I didn't know what had happened but I ran like a bat out of hell.

"When I got up there, Terry was hanging his head out the window and I was yelling, 'Are you all right? Are you all right?' Terry got out of his truck hacking and coughing and waiving his hands around. He never heard me. I had to ask him a couple of more times if he was okay.

"Finally he tells me he's okay and is yelling really loud. I guess the explosion must have made him deaf for a while. I asked him what was wrong and he said, 'When I turned to go up the street, the gun started to fall over so I grabbed it. I guess I squeezed the trigger and it went off.' I talked to him this morning and found out that it's going to cost him $250.00 to fix the hole in the floorboard and another $850.00 to fix the drive shaft."

Everyone in the room was laughing. I went to the office and got a picture of a 1977 Chevrolet pickup truck, one that looked as close to Terry's truck as I could. I then made up a flyer that I sent copies to all of the stations that said, "Will hunt anything and kill it." I put pictures of cans, bottles, his truck, a gas grill that he shot once, and pictures of rats and mice. Each item had a price beside it and I priced the truck at $1,100.00. Once I sent the picture out, I called a few people and let them in on the story.

Everyone I told laughed like hell about it and said they would be more than happy to post the picture in their stations. I don't know what kind of reaction Terry will have to it but my guess is he will not be very happy. He wasn't happy when I sent another picture around because of another incident he had on a hunting trip.

He and a friend of his went deer hunting five or six years ago. They were out there, cold and wet, for two days and didn't see a single deer. On the way back, Terry was riding in his friend's truck when they saw a big, beautiful buck standing on the side of the road. Terry got so excited about it that he grabbed his gun off of the gun rack in the back window, pointed it at the deer and fired a shot. The problem is, he didn't roll the window down before he shot. He blew out the passenger side window, the windshield and the post between the two before he knew what he was doing. And to top it off, he missed the buck completely. After the incident, I took a picture of his friend's truck and sent a flyer to all of the stations with the caption, "Friends Don't Let Friends Hunt Deer Drunk." Then I typed the entire story beneath the picture and caption. He was pissed at me for three weeks after that one. Of course, all I had to do to keep it going was tape pictures of deer all over the station. This time, I printed off about fifty pictures of 1977 Chevrolet pickup trucks and I am going to take them to Station 8 tomorrow so that one of the guys there can keep things going.

We actually had a slow day; not too bad for a Wednesday. I don't mind the slow days too much as long as there aren't many of them. We had a class this afternoon on streets for the rookies and they are coming along nicely. Steve Williams is starting to remember his streets better. Rick is doing great. He'll be relief driving pretty soon. Steve will take a little longer, but he'll get there. This time he at least showed that he is trying to learn his streets. It takes awhile to learn them, and we realize that. Station 18 has the largest first call territories in the city and it takes us a little longer to memorize the streets. Of course, veterans sometimes have trouble remembering them all. I don't know how many times a call has come in and we have had to look it up in our street book. But that's why we made the book. We did it in case of cerebral flatulence, more commonly known as brain fart.

After the class, Steve Buntz came by the station to shoot the breeze. He's a police officer with 25 years of service to the City of Mannington. He comes by every now and then and buys a cup of coffee. Mainly, he's there hiding out so that he will not have to do crossing guard duty. He and Mike Berrier have a running game playing tricks on each other. They try to see who can play the best joke on the other. This particular day, Steve pulled up and came through the engine room and saw Mike. He told Mike, "Wash the windows and check the oil." Then he went inside.

Mike went into the storage closet off the side of the engine room and came out with a small bottle. Mike cleaned his windshield for him. The problem with the cleaning job was what Mike used to clean the windshield. We have a mixture of silicone and water that we call Gorilla Snot. We use it to make the tires look better and it's cheaper to mix ourselves than to go out and buy it. Mike cleaned every square inch of that windshield with the Gorilla Snot.

Steve finished his coffee and walked out to the engine room. He saw that his windshield had been cleaned. He tossed Mike a quarter and said thanks. Then he got in the car and left. A thunderstorm blew up that afternoon and when Steve hit his windshield wipers, all of that Gorilla Snot smeared, blinding him so that he couldn't see where he was going. He had to get out of his cruiser in the middle of a downpour and clean the windshield again.

Afterwards, he came to the station and walked in the door. I was standing at the watch desk and saw him come in dripping wet. He asked me, "Where the hell is Berrier?"

Judging by the tone of his voice, I wasn't about to mess with him. All I did was point down the hall and said, "He just went to the bathroom."

Steve walked down the hall and into the bathroom. I heard a crackling noise and a loud scream. Seconds later, Steve came out of the bathroom laughing like a mad man and putting his Tazer gun back in it's holster. He looked at me and said, "I never thought I would ever see a man try to crawl through a urinal." He walked back to his cruiser, got in and left. I went back to the bathroom and found Mike getting up off of the floor. I asked him, "What the hell happened?"

"That motherfucker shocked me. He actually took his Tazer gun out and shocked me with it."

I looked at him as he stood there. The front of his pants was wet and he was still shaking. I pointed at his pants and said, "That must have been a real pisser."

He looked down and screamed, "That motherfucker made me piss all over myself." I could barely make it down the hall because I from laughing so hard. The other guys came out of the classroom and wanted to know what was going on. All I could do is laugh and point down the hall at the bathroom.

That evening, we went to get groceries and came back with a box of depends and a ton of coupons. I was able to sneak them into his locker and later that evening he came running down the hall yelling that we were real fucking funny. He went out to the engine room and was sitting there pouting.

We went back to the bathroom and took the box of depends out of the trashcan. We all put a pair on over our uniforms and walked out to the engine room. Mike looked up and said, "Y'all think you're real fucking funny. You don't know how bad those burns look where he got me with that Tazer gun."

I said, "Well, we're just trying to make you feel a little less out of place. I know that there aren't many firefighters that wear depends and we just wanted to make you feel a little better."

"Fuck all of you ass holes!" He stormed outside and was going to sit on the bench when he saw his car. I didn't know about this, but Bullet had gone out to his car and put coupons all over it. He came back in and went to the classroom to find that there were coupons stapled from corner to corner on the bulletin board. No matter where he went, he always found those coupons put up everywhere. We had a call later that evening and when he got on the truck, he found coupons taped up inside the cab of the ladder truck. Needless to say, he was extremely pissed off.

As I said, we had a call that evening. The call came in as a working house fire. When we arrived on the scene we found the family standing out in the front yard. Fire was coming out of one front window and a window on the side of the house. There was smoke throughout the house but not a lot of heat, and the fire at that point was contained to one room. As I got off the truck, the owner of the house came over and said that everyone was out of the house and accounted for. Rick and I put our air-packs on and prepared to enter the building. Engine 15 arrived as we were getting ready and I told them to stand by as the rapid intervention team. Every time we go into a house fire, there has to be one man outside the structure for every man inside the structure just in case anything happens. Chief Williams arrived on the scene and assumed command from me.

Rick and I went inside and crawled into the room just to the left of the front door. We could see the seat of the fire and I took the nozzle and popped the flames coming across the ceiling. As I did that, the room went dark for a moment. Rick and I hunkered down and waited for the steam to bank down on us. Once the steam passed, we made our way into the room and headed toward the seat of the fire. I hit the seat and worked the nozzle in a clockwise motion. If I had gone the other direction it would have pulled the heat and smoke toward us. By rotating the nozzle in a clockwise motion, it pulled the heat and smoke away from us.

After the first time I hit the fire, it flared up a little and I hit it again, this time in a wider pattern. We couldn't see anything in there at this point due to the smoke and steam and I didn't notice that I hit the television set. When the cold water hit the television screen, the difference in temperature cooled the screen too fast. The screen exploded and shot glass out in every direction. As soon as it exploded, Rick and I hit the deck and turned away from the direction of the blast as fast as we could. We were showered with chunks of glass, most of which hit us in our backs. It scared the shit out of us, but neither of us was hurt by the flying glass.

After the set exploded, we moved in closer and finished knocking that fire out. The ladder crew set up the ventilation fan and we cleared the house of smoke. We then checked the walls, ceiling and floor for heat. Engine 15's crew went into the crawl space and found no fire extension down there. The squad crew went into the attic and found no damage there, either. We tore out the drywall in the living room and found that it had held up to the fire. We wanted to make sure there was no extension in the walls. The drywall had to be replaced anyway, so basically we were not only tearing it out, we were saving the homeowner a little money because the damaged drywall was gone. On the walls that were only smoke damaged, we used a heat detector to see if there were any hot spots and found none.

It took another hour to complete our salvage and overhaul. After that, we packed up our hose and headed back to the station to clean up. After reloading the truck and taking care of our air-packs, we got cleaned up and sat down to watch the news. I mentioned to Rick that I didn't know what the hell was going on when that television set blew up. He said, "I know what you mean. I didn't know what the hell was something going on."

I said, "Yeah. Scared the hell out of me. I've never had one that blew that big before."

"I'd like to know how often that happens because I'd like to know if other screens blow out like this."

"Well I'll make a phone call to a friend of mine. I don't know if he can help, but I'll see what he says." He knew what I wanted to know about picture tubes. Television picture tubes are vacuum tubes and he told me that some television sets have more pressure inside them than others do. He sent me a list of sets that have higher pressures than others via e-mail. I figured this information would be good for a class the next time we worked.

After getting this information, I went in and watched the news before going to bed. I called Teri and asked about her day. I found out what we had in the mail, a couple of messages left on our answering machine, and what the kids thought of their new teachers. I told her I love her and told her goodnight. I hated that I wasn't there with her but I knew she understands. That's one of the things I love about her. I know she worries about me when I am at work but she wouldn't ask me to do any other kind of work. I was a firefighter when I met her and she knows I love my job. She wouldn't ask me to do anything else.

As I was going to bed, a car pulled up in the back driveway. A couple and their son got out and came in the station. The kid had a lock washer on his finger and couldn't get it off. A lock washer is not a solid circle; it has a slit in one side of it. I called fire communications and told them what we had and that we would be out of service for a few minutes with the call. It was on there pretty tight and could not be pulled off so I went and got the toolbox. When I came back the boy asked me, "Will you have to cut it off?"

I put my hand on his shoulder and said, "Don't worry, bud, I don't think we'll have to do any cutting."

I took a pair of pliers from the toolbox and the pair of Gerber pliers that I wear on my hip and grabbed both sides of the lock washer. It was an old, rusted washer and I was able to pry it open enough to slip it off his finger. I checked the skin around his finger and saw that it was only swollen a little but not bleeding. I told him he shouldn't to do that again and he agreed. They left and I checked us back in service, did the report and went on to bed.

The night was uneventful and we actually got a full night's sleep. The next morning, we sat around talking over coffee and did our shift change. We swapped out with the next shift and I headed over to Station 8 to take the fliers to their A-shift captain. He loved them and started posting them all over the place. Then I headed to my part-time job and to another lovely day in paradise with good old Dick.

When I got to work, Dick was waiting on me with a real shit job. A couple had brought their couch in to have it reupholstered and it was a complete mess. They had six cats and these cats had clawed, shit and pissed all over it. I didn't really want this job and I was ready to quit but I wouldn't let Dick get the better of me. I went ahead and started the job. Dick went out to "run some errands." What he was doing was going down to the corner bar and hanging out there. But I diligently did my work and peeled off all of the layers of stinking, smelling upholstery. I then took every bit of the padding and upholstery up to Dick's office and piled it all on his desk.

He came back that afternoon and hit the roof when he saw what I had done. He started ranting and raving. Going on about how I had ruined his desk and how the stench in his office was unbearable. I just sat there listening and then reminded him that he also worked in this furniture refinishing shop. If his desk was that bad he could refinish it. Then I handed him the price list and walked away. Dick just stood there with his mouth hanging open as I walked away. I clocked out and headed over to get the kids.

The YMCA has an after school program that the kids like. They enjoy hanging out there and it keeps them out of trouble. The kids are great to hang out with and Joey and Nikki have made a lot of friends. When I got there, I saw Joey on one of the outdoor basketball courts having a great time. He ran over to me and said, "Dad, we only have a couple of minutes left in the game. Can I finish the game?"

I said, "Yeah, go back and finish. I'm going to go find your sister."

"Nikki's inside in the craft room."

"Okay. Go have fun."

I went inside and found Nikki drawing a picture. She loves to draw and was working on a picture of a dog. She was doing a great job. She's getting pretty good at drawing and can draw just about anything that she sets her mind to. Her only problem is that she gets down on herself. She is her worst critic. When she saw me, she showed me her picture. I thought it was great. She was a little down on herself because she didn't think the nose looked just right. I thought it looked good.

When she got her things together we went back outside to watch Joey. When his team won the game he gathered his things and then we went home. I got home to find Bullet waiting in the driveway for me. The kids love Bullet. They ran up to him and gave him a big hug. I sent the kids on into the house and asked Bullet, "So, what brings you around?"

He said, "I need to talk to you about something."

"Well, let's go to the deck. We can talk over a couple of beers." When we got out on the deck, we sat down and Bullet just sat there looking at his feet. I finally had to ask, "What's up?"

"You know I love Jennifer, right?"

"Yeah. I know."

"Well, I think about her all of the time. I want to be with her all of the time. I want to ask her to marry me."

"Well, I think she's the right one for you. There's no doubt in my mind about that."

"I think so, too. But what if she doesn't want to get married?"

"Well, it never hurts to ask. And besides, what if you don't ask but she would have said yes?"

We sat there for a little while, talking and drinking beers. When Teri came home, I went inside while he asked her what she thought. When I came back, Teri was telling him that she thought it was a good idea. "Besides, you'll never find anyone else out there that is more perfectly matched to you than Jennifer."

"I know what it is," I said. "He's worried that she shoots better than he does."

Bullet jumped up and shouted, "That's a load of shit." Then, in a weak voice said, "I already know she shoots better than I do." We all laughed about that and then Bullet said, "I just don't know if she's ready to get married. Especially to a man like me."

"Well, I know that you two would be perfect for each other. Teri knows it. Now all you have to do is convince yourself of that."

Bullet hung around for dinner and then went home sounding more confused than he was when he got there. Poor Bullet. He and Nikki have a lot in common. They both get down on themselves so much that after awhile you can't convince them otherwise. Maybe after he's had some time to think about it he'll believe that they are as good for each other as much as Teri and I do.

The next day, I spent the day sick in bed. I hate being sick. I ended up sleeping most of the day, which isn't bad. But every time I woke up I felt like shit. Teri kept calling to make sure I was okay and I would tell her I was fine. Then twenty or thirty minutes later she would call back to check up on me again. I ended up sleeping the day away. She picked the kids up after she got off work. That night I stayed in bed and slept a little more. The next day I felt a lot better and decided to go to work.

Chapter Twelve

We got some bad news today. A fellow firefighter, Tim Simpson, killed himself yesterday. He was a twenty-year veteran of the fire department and had made it to the rank of captain. He leaves behind a wife and two little girls. I never dealt with him very much. He would come by the station once in awhile to drink a cup of coffee and shoot the breeze but not much else.

As the story unfolded, Tim had been out drinking yesterday afternoon and decided to head home. On the way, a man on a bicycle cut him off and he had to slam on his brakes to keep from running over him. When he blew the horn, the cyclist gave him the finger. Tim was known for a very short temper and when the guy flipped him off it pissed Tim off. He pulled into the nearest parking lot and waited for the cyclist to come by.

When the guy came pedaling by, Tim grabbed him off his bike and proceeded to beat the hell out of him. When it was over, Tim was arrested and booked on assault charges. He had to spend several hours in jail because his wife wanted to get some things packed and take the kids with her to her parents' house. That way Tim would have a chance to cool off.

When she finally bailed him out, Tim went home to an empty house. There was no note from his wife. All he saw was that she had packed her things, took the kids, and left him. I guess Tim felt that he didn't have anything to live for. He had been in trouble before because of his temper. He had been in previous fights and was told that one more might spell the end of his career. He thought that he had lost his family. He felt that his career was over. He ended up taking a gun, holding it in front of his chest and fired the shot that took his life. His wife came home this morning and found him, dead.

No one I talked to knew whether Tim called anyone before he killed himself or not. If he did, it didn't do any good. Tim must have felt that he had lost everything. He must have felt that he could not face the consequences of his actions and decided to end it all. I don't know what it would take to make a person take his or her own life. It's a mystery to me. I don't think I could ever do that to Teri, Joey and Nikki. The thought of them suffering as a result of my actions makes me sick to my stomach.

Everyone in the station was in shock over the news. It took all of us by surprise. Some became angry with Tim for "taking the easy way out." Others wondered what his wife and kids would do. The one feeling that we all had was a feeling of loss. Tim was a good firefighter and the department would feel the loss in a lot of ways. Tim was a good friend to a lot of guys and they would feel the loss even more. I didn't know him that well but I felt that I had lost a friend as well.

After we talked about it for a while, I went to call Teri. After telling her what happened, she asked, "Are you okay? How are the other guys handling it?"

I said, "We're doing okay I guess. I didn't know him that well."

"Have you talked to his family yet?"

"No. Bullet said he would try to call Tim's wife later to see if she needs anything. Bullet worked for him for a couple of years."

"Oh, my. How's Bullet handling it?"

"How else would Bullet handle it? He's pissed."

We talked about it a little more and then the tones went off. I gave Teri my love and she told me to be careful. Then I headed to the truck.

We were dispatched to an unknown rescue call. Engine 18, Ladder 18 and Squad 18 wound their way through Saturday afternoon traffic to the scene. We arrived to find a woman and her daughter standing by the side of the road. I got off the truck and asked, "What's the problem, ma'am?"

She was near tears, but pointed up in a tree and said, "My daughter's cat is stuck in that tree and we can't get her out."

"Do what?" After hearing about Tim, I really didn't need this.

"Our cat's stuck in that tree."

"Well, ma'am, what do you want us to do about it?"

"Get her out."

"The fire department doesn't get cats out of trees anymore, ma'am. We don't want to take a chance on getting someone hurt over a cat."

"You have to do it!" I could tell that she was getting angry.

"Ma'am, we don't do that anymore." I then got on the radio and returned the ladder and squad to service. I walked over to the daughter and said, "Don't worry. Your cat will come down when she gets hungry."

The mother stormed over to me and grabbed my arm. She spun me around and said, "Now listen. I pay taxes so that means that I pay your salary. Now I am ordering you to get that goddamned cat out of that fucking tree this instant!"

I was a little shocked at her language in front of her daughter. I removed her hand from my arm and said, "Ma'am, how many cat skeletons have you ever seen in trees?"

She didn't have an answer for that and I got back on the truck. I then stuck my head out window and said, "By the way, since you pay our salaries, I would like to request a raise." With that we headed back to the station. I hated to be such a smart-ass with her but she would not listen to reason. But I will be damned if I am going to risk getting one of my men hurt over a damned cat. Not to mention the fact that it's not exactly safe for the cat.

Usually they end up moving higher into the tree or they end up jumping and getting hurt.

I found out that the service for Tim would be Wednesday morning. I'm off Wednesday. But in a way I wished like hell that I had to work. I hate funerals with a passion. It's bad enough to go to a funeral for someone who dies of natural causes or dies because of an accident. This would be the first funeral I have ever been to because the deceased took their own life. I wasn't sure how I would handle it.

The rest of the day was pretty quiet. We talked about Tim every once in awhile. Bullet finally worked up the nerve to call Tim's wife to see if they needed anything. She wasn't taking calls but he was able to talk to her father. He told her father to pass on the message that if they needed anything to call one of the stations. We wanted to do what we could to help the family. We found out that the funeral would be a fire department funeral. The church service was to be held in a church in Mannington but the graveside service would be in a graveyard in Grandview, thirty minutes away. We started digging out our dress uniforms and checking to see what fit and what didn't fit. Most of us were okay. All we needed were the black bands for our badges.

The tones went off and we were sent on an unknown medical call. When we arrived, we found a woman standing out in her driveway waiting for us. I asked her what the problem was. She said, "You gotta help my husband. He's got smiling mighty Jesus." I didn't know what to think of that one. I didn't know if he had the shits or if there was something seriously wrong with him. I asked her again what the problem was and she said, "My husband has smiling mighty Jesus."

The ambulance arrived and I had one of the paramedics go talk to her. He listened to what she had to say and had no idea what the hell she was talking about.

Then he had her describe her husband's symptoms. When she finished, he said, "Do you mean spinal meningitis?"

"That's what I said, smiling mighty Jesus."

Once we knew what the man had, we took the necessary precautions and went in to treat him. He was having a bad night and wanted to go to the hospital to be checked out. There wasn't much to do but package him up on the stretcher and take him to the truck. We headed back to the station and I did the report and went to bed. I didn't sleep much because I was thinking about Tim.

The next morning, we sat quietly around the table, no one saying much. We didn't have that many calls the day before so we just went through our shift change. Today is Sunday and I don't have to worry about working part-time. I headed to the house and just laid around. I didn't go to church with the family. I just wanted to be alone for a while.

That afternoon I spent most of my time in my studio playing the piano and lost in my thoughts. I was just rambling through songs, not really concentrating on what I was playing. I kept thinking about Tim and how he could do that to his family. I couldn't understand why he would do it. I know that I love my job just like Tim did. He would have probably lost his job, but that's not the end of the world. If I lost my job, I don't think I would end my life over it. I would find something else to do with my time. There is no need to kill yourself over losing a job.

I have seen people who have had to go out on disability who fought like hell to keep the job they were being forced out of. None of those guys ever committed suicide because they couldn't do their job anymore. I can't understand why someone like Tim, who had a college degree and could have done any number of jobs, would kill himself. I know that most firefighters don't feel that they do this as a job, but the career is who they are. I don't feel that I work as a firefighter, I feel like I am a firefighter. It's a big part of who I am, not what I do. I worked in television before becoming a firefighter; I was a cameraman for several different television stations. But I always felt that I worked as a cameraman, I never felt that my job was a part of me like being a fire fighting is.

The people who have been forced to go out on disability didn't want to go out partly because of a love for the career and partly because being a firefighter was a big part of them. But they never once considered taking their own life because of losing a part of their identity. I don't think I could ever do that. Not only to myself but also to my family. I don't claim that I am the be-all-end-all of this family. But I think that it would devastate my family if I even attempted suicide. I could never do that to them. I just know that I would be able to find something else to do for a living. Besides, I'm not sure I would have the guts to do that. I don't think I could kill someone else, much less myself.

I didn't sleep well that night either. I kept thinking about Tim and how his family was doing. I was wondering what they were thinking and wondering how they would make it since Tim was gone. That's another reason that I couldn't do that to Teri. She means the world to me and I would never do anything to intentionally hurt her. I know that with her help, I would be able to make it through anything. No matter how bad things got. I would still be Dalton James. Husband. Father. King of my own castle. I know that with the help of my family I would be able to get through anything. After coming to that conclusion, I fell asleep.

The next day, I went to work at the part-time job and had a fairly quiet day. Old Dick was out sick for the day so I didn't have to put up with his shit. I did go to his office and mess with a few things that I knew would piss him off. That made me feel a little better. I didn't do anything major, I just put shaving cream on the ear piece of his telephone, glued his chair three feet from his desk, and mixed his files up. After that I went to work reupholstering a couch.

After work, I picked the kids up and stopped off for burgers on the way home. We had dinner and as we were cleaning up, Teri asked, "How are you holding up?"

I said, "Right now, fine. I just hope everything goes okay Wednesday morning. That's going to be the hard part. I hope it goes quickly and I don't have to dwell on it very much."

"Do you want me to go with you?"

"You don't need to worry about that. If you're going to take a day off, I would rather you do it for fun. Not this."

"Well, if you change your mind, let me know. I don't like funerals any better than you do but if you want me there for support, just ask."

"Thank you, baby. Just knowing that will help."

I turned in early that night. I hadn't slept much the last two nights and I was exhausted. I quickly fell asleep and slept fairly well. The only time I woke up was when Teri came to bed. I moved closer to her and went back to sleep holding her in my arms. The next morning, I kissed everyone, gave them my love, and headed to work.

Chapter Thirteen

The day started off as usual. Everyone was getting over the shock of Tim Simpson's suicide and things were getting back to normal. I checked around with the guys on A-shift that morning to see if anyone needed me to work in their place so they could go to the funeral. Tim worked on A-shift and I wanted to make sure all of the guys that worked with him would have a chance to go. Everyone had made arrangements to get time off so no one needed me to work for them. I was glad that they would get to go to the funeral but I was disappointed that I didn't have to work. I had already made arrangements with Rick Johnson, my boss for my part-time job, to get the day off.

I really didn't want to go to the funeral. It's not that I didn't like Tim or that I was pissed off at him for killing himself. I just hate funerals, no matter who they are for. That and I didn't really want to see a fellow firefighter laid to rest. It didn't matter that he killed himself; I wouldn't have wanted to go even if he had died in the line of duty or of natural causes. But I would go and pay my respects. I didn't want to. But I would go and be there for the other firefighters.

The first call of the day was a reported house fire. We arrived to find that one of the back rooms of the house was burning and it appeared to be moving out into the hallway. As we made our way into the house, I saw fire at the end of the hall. I opened the nozzle and hit the fire but nothing happened. I thought that maybe we weren't reaching the end of the hall. It was a huge house and a very long hallway. We moved closer to the fire and I hit it again. Again nothing happened. I started to wonder just what was burning when I noticed that there was a room behind the fire. I started trying to figure out what kind of floor plan this house had as we moved closer to the fire.

When I got close enough, I finally figured it out. The "fire" that I was hitting was a reflection in a mirror. I couldn't believe it. All this time I thought I was hitting the seat of the fire and I had been hitting a reflection the whole time. The mirror was mounted on a closet door and we could not see the doorknob because of the smoke. The door was open just enough so that the reflection was of the room instead of the hallway. We never hit the fire until we made our way down the hall and turned the corner into the fire room. I felt like a complete idiot because I couldn't figure that out in the beginning.

Once we got to the fire room, we knocked the fire out in only a few minutes. There wasn't a lot of damage to the house other than that particular room. Of course, there was a little more water damage than necessary, but nobody said a thing about that. We overhauled the room and salvaged what we could. There wasn't much to salvage, only a couple of things on the floor. We tore out the walls on the burned side so that we could check for fire extension and took care of the debris. Once that was accomplished, the fire investigators came in and determined that the fire was caused because of a bad electrical outlet.

After the fire, we went back to the house and cleaned up. I took a little ribbing about fighting fires in mirrors but Rick helped me convince them that you could not really tell where the fire was. I did find a little mirror on my desk later that afternoon. I took the jokes in stride and didn't let them get to me. That's one of the best defense mechanisms around here. If you don't let the things they say get to you, they will eventually shut up.

There have been other fires when someone has a surprise like that. There have been cases when other men have gone in to fight a fire and ended up only hitting a mirror. I was on a call once where there weren't enough men on the call to help out. For every man you have inside the house, there needs to be a man on the outside as well. When the first due engine arrived, the captain decided to hit the fire from the window of the house. They popped the window and shoved the nozzle inside and worked it in a circular motion trying to knock the fire down. They caused a lot of steam but every time the steam lifted the fire was still burning. They couldn't figure it out until the second unit came in. They were attacking the fire from the side of the house and the only light was from the fire itself. Once they opened the door on the side of the house, they went in and found that there was a solid glass wall from one end of the room to the other. The house was being remodeled and the old glass porch had not been removed yet. Every time they thought they were hitting the fire, they were hitting the glass and never reaching the fire. I'm just surprised that the glass didn't explode when the cold water hit it but miraculously it didn't. If it had, it would have hurt those guys pretty bad.

My first fire was to an old house. We crawled almost all the way through the house and finally saw a glow in the corner. We hit what we thought was the seat of the fire but when the heat and smoke never let up, we moved in closer to find that all we had put out was an old kerosene heater in the corner. We had to go further into the house to finally find the seat of the fire. We put it out and there were jokes going left and right about us putting the heater out. It happens a lot. But we make sure the fire is out before we leave the scene.

That afternoon there was a bad thunderstorm. Alarms started coming in from all over town. Every single alarm was false. We rode twenty-seven false alarms in an hour and a half. When the alarms start coming in that quickly, the department goes into what we call the storm plan. That means that the only one to ride an alarm call is the first due engine. I don't know how many the entire department rode, but Engine 18 rode a total of twenty-seven. I didn't get dinner until around ten o'clock.

The last false alarm was to one of the worst neighborhoods in town. This is an area where there are murders every day, drug deals by the truck load, hookers on the sidewalk at all hours, and the bars are the type of places where they sweep the eyeballs up after closing. We checked out the business where the alarm came from and of course found nothing wrong. We got back on the truck and started to head for the house when Bullet slammed on the brakes. I looked at him and started to ask what was going on but he bailed out of the truck. I saw that he was heading for the rear of the truck and went after him. When I got back there, I saw Bullet body slam a man. I could see that there was another man lying in the street about fifteen feet from the tailboard of the truck. I looked at Bullet and said, "Bullet, what the hell are you doing?"

Bullet looked at the guy and said, "Mister, don't you know better than to try to hitchhike with us? We ain't no taxicab. No riders!!"

I found out that Bullet had looked in the mirror and saw a man running towards the truck. Just when he jumped for the tailboard, Bullet slammed on the brakes. That man bounced off the back of the truck and he was the laying fifteen feet from the tailboard. When Bullet got to the back of the truck to check on him, he looked up and saw another man had already jumped on the tailboard. The guy didn't even have sense enough to run when the truck stopped. Bullet grabbed him and body slammed him right there in the middle of the road.

I went to check on the guy that bounced off the truck. He was conscious and alert but his head, shoulder and legs were hurting like hell. I couldn't help but laugh at the thought of seeing him bounce off the truck. Bullet was pissed when we started checking them out. He looked at me and said, "You mean to tell me these guys try to jump on my goddamned truck and now we gotta treat their sorry asses?"

I used as soothing a voice as I could muster and said, "Well, Bullet, that is the business we're in. We have to treat all victims no matter their race, creed, color or level of stupidity."

"Well if this don't beat all. These motherfuckers piss me off and now I gotta be nice to them. That really sucks, Cap."

"Well then why don't you go over there and direct traffic. Then you don't have to deal with these two."

"Gladly."

Bullet started to head over to direct traffic when I said, "Bullet."

"Yeah?"

"Try not to create anymore trouble, will you?"

"Fuck you, Cap."

I laughed and tended to my patient. EMS arrived and hauled both to the hospital with a police escort. And of course the one that bounced off the truck wanted to press charges against Bullet for slamming on the brakes. He was claiming vehicular assault but decided to drop his claim when he was informed that it couldn't have been vehicular assault since Bullet didn't back over him. Bullet said, "I'd be glad to give you a stronger case."

The rest of the night was quiet. I didn't sleep to well knowing what was in store for us the next morning. I was dreading the funeral and I hadn't even started putting my dress uniform on. I laid it out and got all the hardware on it. I had to take the badge off because I forgot the black band but I did get everything else right. I laid awake that night trying to figure out what I would say to the family if I had to talk to them. I'm never good at things like that and I was dreading it the most. I figured that I would just say how sorry I was and then try to find some other place to be. Hopefully I would be able to avoid talking to anyone. I hated to be that way but I never know what to say or if what I am saying is appropriate. I did get a little sleep but it was never a restful sleep.

The next morning, we went through our shift change. Then I got in my dress uniform and headed to the church. When I arrived, a lot of the department was already there. We were all standing in back of the church, fidgeting with our ties and mumbling about how we hated those uniforms. The only time I liked wearing my dress uniform was the day I got married. Instead of a tuxedo or a suit, I wore my dress uniform. All of my ushers were firefighters and they wore their dress uniforms, too. I liked the dress uniform that day but, even though I hadn't gained any weight since the day I was married, the dress uniform felt like it was tight as hell.

We stood around out there as long as we could. None of us wanted to go into the church. I would have rather gone to my part-time job and kissed Old Dick's ass all day long than go into that church. I didn't want to see the casket knowing who was in there. I didn't want to think about losing a fellow firefighter. I didn't want to see the grieving widow or the two little girls that had just lost their daddy. I didn't want to deal with any of this. I just wanted to get the hell out of there but I knew that I had to go. I don't know what it is about funerals that I hate so much but I just cannot stand to be at them.

I didn't hear much of the eulogy or what the preacher had to say. I just sat there looking at my hands and I was thinking about how I really didn't want to be there. I couldn't figure out why I hated funerals so much. I deal with death and dying people a lot on the job. I have carried dead people out of houses throughout my career. I have dealt with dead babies, dead adults, dead animals, you name it, and I have had to carry all of them out. I feel for the family's loss but it never affects me like a funeral does. I guess it's the fact that it isn't really final until the casket goes into the ground. I thought to myself that I hope I go before Teri does. If a funeral for an acquaintance is this hard on me, I have no idea what I would do if I ever lost her. Probably crawl into the casket and beg them to let me go with her.

Finally, that part of the service was over and we headed out to the parking lot to drive to Grandview. We had to wait until the honor guard carried the casket out of the church and to the fire truck that would carry Tim to the graveyard. Tim was the A-shift captain of engine 16 and the men had off-loaded the hose. They had draped the truck in black on both sides and the casket was covered with the American flag. Behind Engine 16 was a line of more than 150 vehicles loaded with family and friends of Tim. I didn't want to go to the graveside funeral, but I figured I had gone this far, I would go all the way with it.

As we drove to the graveyard, I cranked up the stereo and let a little AC/DC take me away from all of this. "Back in Black" was on the stereo and it was making me feel a little better. I didn't think about it until later but "Back in Black" probably wasn't very appropriate at the time. But I guess it beats another AC/DC song, "Highway to Hell."

As we wound our way through the two towns, each time we passed a fire station, the crew was standing at attention on the front ramp, flag at half-mast and the truck was parked out front with black ribbon draped on it. When we went through an intersection, there was a police officer standing at ease with his head bowed and his hat over his heart. I thought this was a nice gesture, especially in Grandview where they didn't even know Tim. It was a last farewell to him.

We finally arrived at the graveside and started assembling for the service. I couldn't hear much of what was said by the preacher because he was so far away. The coffin was sitting in front of me and all I could do was stare at it. I also kept looking at Tim's widow and seeing her cry. It was heartbreaking. I never want to go through that for any reason. The only problem is that I never want Teri to have to go through that either. Maybe I should just hope that we live forever and neither of us has to go through it or, since we can't live forever, that we die together. I know one thing is for sure, if Teri was to go before me, I might as well be dead because life would not be worth living.

I did okay throughout the service. The bagpipes were getting to me a little as the lone piper played "Amazing Grace." I have never had a problem listening to bagpipe music before but on this particular occasion I didn't care to hear them. I did okay when we stood at attention and saluted as the honor guard folded the flag draped over Tim's casket and when the captain of the honor guard handed the flag to Tim's wife. Then she took her daughters to the casket and they each laid a flower on top of it. When her youngest daughter, three years old, laid her flower on the casket, she said, "Bye, bye, daddy."

That's when I lost it. I couldn't help it. She said it in such a sweet way and it was as if she expected to see him come home from work the next morning. I felt a rush of emotions all at once. I felt sorry for Tim's wife who had lost her husband and I felt sorry for his daughters who had lost their father. I felt sorry for his youngest daughter because she wasn't old enough to understand that her daddy wasn't coming home again. She said goodbye to her father but I wasn't sure if she truly understood why she was saying goodbye. All I knew was that I didn't want to see my family have to go through the things this family was going through. It just broke my heart to hear her say, "Bye, bye, daddy."

As I looked around after we were dismissed from the service, there wasn't a dry eye in the crowd. Most of the guys I talked to lost it when she said goodbye. I looked around and saw these firefighters, these men with true grit, who deal with life and death on a daily basis, and we all were crying for a little girl who had lost her daddy. It just took me by surprise.

After the service, I headed over to the hospital to see Teri. I just had to see her face and get a hug from her. I had to see her to help me relax. I didn't want to be away from her at that point. I just had to see her. When I walked up to the desk in the ER, she came out of one of the rooms and saw me. She came over, looked me in the eye and asked, "Are you okay?"

I said, "I was until I heard his three year old daughter say, 'Bye, bye, daddy.'"

With that, she gave me a hug and held onto me for a few minutes. Then I wiped my eyes and looked her in the eye and said, "I need you to promise me one thing."

"What's that?"

"Don't let me get that bad. Don't let me get desperate enough to attempt suicide. I don't ever want you and the kids to ever have to go through that. Just promise me you won't let me get that bad."

She hugged me tight and said, "I promise, baby. I promise." We held each other for a few more minutes and then I left and headed home.

I stopped by the station on the way home to get out of that uniform. I didn't want to wear it anymore. I just wanted to get into some shorts and a T-shirt and relax for the rest of the day. I didn't want to deal with any of this today. Today, I just wanted to be me, Dalton. Not a captain, not a firefighter, just me. I went in the studio and stayed there until it was time to go get the kids. After picking them up, we went home and I lay around the house watching television and hoping that these thoughts would get the hell out of my head.

We had dinner and being with the family made me feel a little better. I was starting to feel like myself again and that's when I realized that I would get through this. I had lost a friend and a fellow firefighter but I would get through this and get on with my life. Being with the family helped me realize that.

I did, however, come up with my ideal funeral. It definitely isn't your typical service. I didn't want a graveside service or a church service. I told Teri that I want my wake held at a strip joint. I want people to have fun and enjoy themselves. There would be no kids, just adults, and I would have a flat, glass-topped casket. There are to be strippers dancing on top of my casket and, if possible, I want my eyes opened so that I can see them, too. I then want to be cremated. Let the fire win for once. Then my ashes are to be sprinkled over the ocean so I can get in a woman's panties one more time.

I realize that this is a wish that will probably never come true but what the hell. It would be nice to go out seeing naked women dancing over me. And, who knows, people might even have a good time. I just don't want people crying over me. I want them to have a good time and enjoy the show. Maybe we could have male strippers for the ladies. Just keep their asses off my casket.

I went to bed that night and slept like a baby. I had a rough night before and needed to get some sleep. Being there with Teri and knowing that my children were in the house with me helped out a great deal. I didn't think of Tim or why he committed suicide once the whole time I was going to sleep. I just slept with Teri in my arms and the feeling of love I have for my family. I guess there was a sense of relief that I no longer had to deal with going to the funeral. I woke the next morning feeling a lot better.

I went into work and Old Dick was there bitching about what had happened in his office on Monday. All I told him was prove that it was me. He couldn't do that and it pissed him off even more. I didn't care. I wasn't going to let him get to me today. I just wanted to do my work and not deal with this asshole. I took my CD player out of my locker and cranked it up as loud as it would go. I did my work and didn't have to listen to his bullshit. I wear the ear buds instead of the regular headphones and Dick couldn't see them. He just thought I was ignoring him. He went and complained to Rick Jackson about my attitude and that I was ignoring everything he said. Rick came over and asked me what the problem was. When I took the ear buds out to hear him, he understood why it looked like I was ignoring Dick and said, "Never mind." He went back to Dick and told him to leave me alone and let me do my job. Of course, that pissed Dick off even more.

    I got off work and got the kids. We picked up Teri and we went out for a night on the town. We went out to dinner and caught an early movie. I knew the kids had finished their homework because they have to do that before they are allowed to play games at the YMCA. The counselors there make sure they do all their homework assignments before doing anything else.

    After dinner and a good movie, we headed home. I sent the kids to bed with a kiss and I headed to bed to do a little bit of cuddling with Teri. All in all, it turned out to be a nice day. I didn't have to deal with Dick that much, I had a good dinner with my three favorite people in the world, saw a good movie and got lucky that night. Pretty nice day, if you ask me.

Chapter Fourteen

Things were getting back to normal at work. The morning and afternoon went slow as hell. Kind of odd for a Friday but I had a feeling that things would pick up soon enough. All we did during the day was station and truck maintenance and a video for afternoon class. Hell, most of us slept through the video but I figured that we picked up enough through osmosis that it still counted as class.

That evening, we sat around watching television until around 10:30 when the tones went off for a vehicle accident with injuries. We arrived on the scene to find an overturned car with power lines down all over the car. The victim was still inside but we could do nothing to help him until we knew for sure that the power lines were dead. I called for the electric company. Mannington has its own electrical co-op and that part of town is both Mannington Electric and the state power company. In that part of town, both power companies have lines and there is no telling who owns the lines if you cannot see the tag on the pole. The tag on this pole gone so I could not tell who owned it. I ended up calling both power companies out to the scene.

While we waited, a bystander came up to the scene and said that he was a doctor. He said that we really needed to get that man out of the car because he needed help desperately. I explained that we could not go to the car until we knew whether the lines were live or not. We had no way of telling if they were live lines and that I would not put any of my men in harms way until we knew for sure. He kept arguing that we had to do something but I wouldn't budge until the power company came to the scene.

EMS arrived on the scene and I went over and explained the situation to them. They agreed to wait for the power company and we stood around until they arrived. As we waited, Rick came to me and asked who the hell the guy was standing next to the car. I looked over and the doctor had made his way over to the side of the vehicle. He was looking inside at the victim. I walked closer to the car and yelled, "Hey, doc, get your ass out of there!"

He turned and yelled, "This man is hurt and he needs help! I made it over here unharmed, I don't see why you can't get him out of here."

"Doc, just get your ass out of there right now. We don't know if these lines are live or not and we don't want another victim. Now, move your ass!"

He reluctantly walked away from the car and back to where we were standing. When he got there, the police took him into custody. That's when one of the paramedics said, "Doctor Barlow?"

The doctor turned and looked at him and said, "Yes, that's right, Doctor Barlow. You really need to help that man."

The police took him away and I turned to the paramedic who knew him. I asked, "What the hell is that man's problem? He could have been killed."

The paramedic said, "He just wanted to help the guy out. I guess he let the situation get the better of him."

"What kind of doctor is he?"

"He's a dentist."

I threw my hands in the air and walked off. Loud enough for the doctor to hear, I said, "Well what the hell was he going to do for the guy, clean his fucking teeth?"

We had to wait for nearly forty-five minutes for anyone from either power company to show up. The first to arrive was the state power company and said he was on call from his home in Stansbury, thirty minutes away. He said that he had heard on the radio that the man coming from Mannington Electric was on his way from Jackson, nearly an hour away. I asked the guy if there was anything he could do to help us. He said that the lines belonged to Mannington and there was nothing that he could do. I asked him, "Well could you at least tell us if the lines are live or not? We've got a victim trapped in that car and we need to get to him."

He looked up at the lines and said, "I'll check them for you but I doubt they're live. Let me take a look and I'll get back to you. He went to check the lines the lines to see if they were hot. He came back and told me that the lines were not charged and that we could safely work around the them with no trouble.

We quickly got the victim out of the car. We loaded him on a backboard and packaged him up to make the trip to the hospital. He was unconscious and unresponsive to us but he was alive. He had a good strong pulse and he was breathing okay. The tow truck had arrived while we were waiting for the power company so we didn't have to wait on him. He hooked the car up and turned it over. Then a rollback came in and took the car away.

Later, we found out that the kid in the car wrecked as he was running from the police. He was clocked at 100 miles per hour when he passed a cop on the side of the road. The officer gave chase but never got to write the ticket. As he neared the kid's vehicle, the kid missed a turn, went through a brick wall and the car flipped. After flipping over, the car slid and clipped the utility pole.

We also found out later that the kid's father was suing the city of Mannington. The suit wasn't against the fire department but against the power company for taking so long. The kid turned out to be paralyzed from the waist down and there was nothing that the doctors could do for him. I don't know how it will turn out but I do know that we did everything that we could for him. I guess Mannington Electric needs to put someone closer on call instead of having them come from such a far distance away.

We headed back to the house and had just got in the door when he tones went off. We had to head out again to a reported house fire. I knew that this was going to be a long night. We arrived to find that the basement of the house fully involved. Fortunately, the basement could be entered from the outside.

When there is no outside entrance, you have to go down the stairs inside the house. The stairwell usually acts like a chimney and you have to go through all of the heat and smoke. It's a miserable, dangerous trip when you have to do that. I went down into a basement once when the basement flashed over as we were heading down the stairs. The flames shot up the stairwell and right into our faces. I was in the front on the nozzle and the heat generated by the flash over was so intense that it melted the regulator off of the face piece of my air-pack. I couldn't go in and fight the fire because I had to go out and get a new air-pack. If it hadn't been for the guy behind me pulling me back up the steps, I'm not sure if I would have made it out of there unharmed.

Rick and I donned our air-packs and took the line to the basement door. Engine 15 arrived and backed us up as we started for the basement. When we raised the garage door of the basement, we immediately started attacking the fire all around a large stack of oak lumber six feet high and wider than the garage door. We tried to work our way around the wood and get in behind it but there wasn't enough room to get by on either side of the stack. We found a standard door on the backside of the house and entered there. As we made our way inside, the owner of the house came running to the door and started yelling, "Forget the damned house. Save the wood!"

I couldn't figure out why he would want to keep the wood and didn't care about the house. We just ignored him and kept doing our job. We got further into the basement and moved around behind the wood to the furnace room. I felt the door and it was as hot as hell. I looked beside the door and saw a vent in the wall that was glowing red. I keep a small, flat pry bar in my turnout gear and I used it to pry the vent off the wall. We stuck the nozzle in the hole and worked it back and forth for a few seconds. Then shut it down. The glow was gone but there was still a lot of smoke and steam coming out so we hit it again.

After the second time the smoke had diminished and the steam wasn't as bad. I was able to look inside and saw an air duct running vertically up the wall. I called for an ax or sledgehammer to knock the vent away. A man from Engine 15's crew came in with a pick-head ax. I knocked the duct back and was hit in the face with a blast of heat. I ducked down and told Rick to shove the nozzle in there and hit it again.

After a few seconds of water, Rick stepped back and I was able to get up there and knock the duct to the floor. Rick hit the fire one more time and then we opened the door. We went inside and knocked the rest of the fire down. We set up some lights and determined that the fire had been contained that room only. The room was completely closed in with drywall ceilings and walls. The floor was concrete and didn't conduct much heat. We tore down the drywall in the room and checked for fire extension into the walls and floor joists. There was a little scorching where the ducts went out of the room but we wet those areas down and there were no other problems.

When we finished our overhaul of the room, I went outside to take a break. I saw the owner of the house standing with his family by the side of the road. I walked up to him and said, "I have ask you, why did you want us to save the stack of wood in the house and not the house itself?"

He said, "Well, the house is insured, the wood ain't." I gave him a confused look and he explained, "I paid over $ 2,000.00 for that lumber. It's oak lumber that I was going to use to build some furniture. I was worried about it since I have a leaky basement so I called my insurance agent. He told me that the wood would not be included in my insurance until I made the furniture. I just figured that since the rest of the stuff in the house was insured, fuck that, save the wood."

I had to laugh, "Well, I gotta tell you, that's a new one on me. But you'll be happy to know that the wood and the rest of the house are fine. Your furnace is history and you'll have to redo the drywall in that room but that's about it."

The family thanked us and as I was walking away I heard the wife jumping on him about being so concerned about "that damned pile of wood." I headed over to the rehab truck and got some water. I had to tell the guys what the owner of the house told me. We all laughed about it and took a break. The fire investigators were in the house looking for the cause. Until they were finished, we sat around waiting and wishing we were in bed asleep and not out here sweating our asses off waiting for the investigators to finish.

After nearly an hour, the investigators finished their work and we went back inside to finish cleaning up the room. We hauled out everything that we tore off of the walls and ceiling and we carried the furnace out of the room. We found out from one investigator that there was a short in the air conditioning system and it lit off some paper that was stuffed in the holes around the ducts.

Apparently, when the room was closed in, the owner was told that if he sealed the room tight, it would make his furnace work more efficiently. He used a spray in foam insulation around the ducts, which would have been fine. The problem is that the holes were fairly large and he had to back the holes with something or else he would have been spraying right through the holes and they would not have been plugged. He packed the holes with newspaper from the inside of the room and sprayed the insulation in from the outside. If he had used the insulation on both sides there probably would not have been a fire. The short lit off some of the newspaper and when it caught fire it fell out from around the air ducts. When the burning paper hit the floor, it lit off sawdust on the floor that had probably been in there for years. The owner had also been using the room as a storage room as well as a furnace room and there were cans of furniture stain, paint thinner, and other flammable liquids in there. After the sawdust caught fire it was just a matter of time before all of those liquids lit off.

We finished cleaning out the room and took our hoses up. We headed back to the station and cleaned everything and reloaded the truck. I called fire communications and returned us to service and hit the showers. I finally got to bed at 3:30 and fell asleep pretty quick. The tones went off at 4:00 sending us to a possible vehicle accident with injuries. As I headed to the truck, I thought to myself, "How the hell do you have a possible vehicle accident. Either there was a crash or there wasn't."

We arrived to find a car in a ditch with the motor still running. Other than being hung up on a manhole cover, there was nothing wrong with the car. I walked over and saw that the driver was unconscious. I knocked on the window but there was no response. I could see that the man was breathing, so we knew he was alive. I looked in the back seat and saw quite a few beer cans back there. Then I knocked on the windows several more times. After knocking on the window the sixth time, the man moved, almost as if he was trying to get away from the noise.

I took Bullet's flashlight and laid it on the driver's side of the hood. Then I took my own flashlight and laid it on the passenger side. I told Bullet to pull the truck up next to the man's car and start blowing the horn. Bullet pulled up to where the front bumper of our truck was equal to the man's door and started blowing the horn. After the third blast from the air horn, the man woke up from his drunken sleep and, seeing the two flashlights on the hood, screamed and grabbed the wheel of the car. We all burst into laughter and I walked to the window of his car and knocked again. This time he rolled the window down.

I had the man turn the motor off and get out of the car. He got out and immediately fell to his knees. When I instructed him to get out, I forgot to tell him that there was a three-foot drop right outside his door. He dropped into the ditch and fell to his knees. When he got up I helped him out of the ditch and into the waiting arms of the police. They took him into custody and had his car towed to the impound lot. We headed back to the station and I did the report. Then I went to take another nap before shift change.

The next morning we told A-shift what we had the day before and what we did to get back into service. They loved the story about the drunk. I think I would have enjoyed telling it more if I wasn't so damned tired. Fortunately, it was Saturday and I didn't have to go in to my part-time job. I drove home and crawled into bed between Teri and the dogs. As I drifted off to sleep I thought to myself, "One of these days we're going to have to get a bigger bed."

I slept for a few hours and decided to get up and join the family. We headed to the mall to do a little shopping and dinner. We didn't buy much. We just walked around window shopping and laughing at all of the teenagers walking around in their baggy pants. We saw a couple of guys picking back and forth. When one took off running, the one he was picking with took off after him. His pants were so baggy that when he went to run, they fell down and he tripped in the middle of the floor. He got up and tried to act cool but didn't realize that his pants were still down. I turned to Teri and said, "It's hard to be tough when your pants are around your ankles."

Before we left, I had to go to the bathroom. I headed down to the end of the mall and did what I needed to do. On the way back, there was a kid standing out in front of a gothic store. I'm not sure what the hell he was advertising but it wasn't a strategic business move. He had a haircut that made him look like Moe Howard of the Three Stooges. Everything he was wearing was black or white. He wore four-inch platform shoes and black and white striped socks. He wore baggy shorts that hung down to the middle of his shins. He had a black tee shirt on with a tuxedo coat with tails on over that. He had top hat on that was way too small to fit him. He was carrying a walking stick and had rings on every finger. Lastly, he had so many piercings on his face that it looked like he fell into a tackle box.

When I walked by this guy, I couldn't help but laugh out loud. When he heard me laughing, he turned and yelled, "What the fuck are you laughing at, ass hole?"

I stopped, turned and looked him in the eye, and said, "I don't know. You tell me what the fuck you are and I'll tell you what the fuck I'm laughing at."

An elderly couple was walking by when I said that and both of them started laughing. It pissed the guy off so much that he turned and stormed back into the gothic store. I didn't mean to laugh at him but he looked so damned ridiculous that I couldn't help it. As I was walking away, he stepped out of the store with a couple of his friends and yelled to me, "I'd like to see you come back here and laugh at me now."

I turned and started heading back in their direction. All three of them chickened out and went back into the store. When they turned and ran, I decided to skip it and go back to Teri and the kids. She was wondering what took me so long and I told her what happened. She loved it. She couldn't believe that I would say that to him but she loved it.

We went home and I crashed early. I was tired and could hardly keep my eyes open. I must have slept pretty soundly because I don't remember Teri coming to bed. All I remember is going to bed and waking up the next morning with the five of us in bed. Teri, the three dogs and me. We ended up playing hooky from church and decided to have a slow, quiet day at home.

It was raining outside and we didn't want to get out into it. We watched the football games on TV and played a few games. I had to work the next day so I wanted to get some time with the family. I could have done some things in the shop but I didn't care to. All I wanted to do was hang out with the family. Not an eventful day, but a good day.

Chapter Fifteen

Jack Walters, from EMS, and his partner stopped by the station this morning. They were coming on duty and decided to come by and see Mark, Jack's brother. They told us about a call they had a couple of nights ago. A young couple had been having some problems. The boy found out that his girlfriend had been cheating on him with his best friend. He kept his cool and didn't let on to her that he knew about the affair.

They had gone out to dinner and when they came home they went straight to the bedroom. The girl apparently had several body piercings, and one thing that she had pierced was her vaginal lips. The boyfriend acted like he wanted her really bad and got her to take the ring from down there. When he moved in for a "closer look" he slipped a small padlock into the pierced holes, locked it and left her there. She didn't know what to do so she called for an ambulance.

When they arrived on the scene, she wanted them to cut or break the lock but Jack said that she didn't look like the type of person that he wanted to be touching in that region. As he put it, "I left my body condom in my other suit." He decided to put her on the stretcher and take her to the emergency room. He left her there and let the ER doctors handle the problem.

We were all laughing and I asked, "Why did you take her to the ER?"

"So they could cut that lock off. I wasn't about to touch her."

"But why didn't you take her to a locksmith?"

We all laughed and then started station cleanup. I went to the office to do some paperwork. I took care of training records that I hadn't had a chance to enter into the computer. Bullet came in the office and we talked about training and what we needed to do in the future. Bullet had been wanting to have a class on vehicle extrication and he got on the phone to a friend of his who owned a junkyard. He arranged to get four cars to cut up and practice on. We also set up a couple of other classes we needed and a couple that were just for days when we didn't feel like learning anything new.

We had just sat down for lunch when the tones went off. We were dispatched to an injury from a fall. As we were en route to the call, fire communications called us and told us to stage at the scene until the police department cleared the scene. I was a little confused because we don't ordinarily stage for injuries from falls. We staged a block away from the address until PD came back with an all clear. When we arrived, we found a four-year-old boy on the front porch with a cut on his forehead and difficulty breathing. When we got off the truck, I put the medical bag on my shoulder and carried the oxygen bottle in my right hand. As I was walking up the steps, the father came busting out the front door heading straight at me. As he was coming at me he yelled, "Motherfucker, you ain't touching that boy." He reached out to grab me so I took the oxygen bottle and popped him in the forehead with it. That raised a big goose egg on his head and he stumbled over to the left of me. I went ahead to the porch while the police came out of the house and took the father into custody.

The father was the one who put the cut on his son's head. His wife told me that her husband got pissed at the boy and grabbed him by both arms, then threw him head first into a wall in the house. She said she grabbed the cordless phone and her son and went out on the porch. She slid a couple of chairs on the porch in front of the door to keep him from coming out after them. Fortunately, he was too fucking stupid to realize he had a back door and didn't go around the house to get at them. She called the police and asked for an ambulance as well.

We bandaged the boy's head and took a look at his ribs. He was complaining about some pain in his right side. I checked to see that the ribs were moving correctly with his breathing. A lot of times there is condition called paradoxical movement, a section of the chest wall moves in when the person inhales and moves out when they exhale. The boy didn't have any paradoxical movement in his chest, just some pain along his right side. We put a cervical collar on his neck and put him on a backboard just as a precaution. We put him on the stretcher and rolled him to the ambulance.

After the ambulance pulled off, one of the police officers, Charlie Jameson, came up to me and asked me to check on the father. He was claiming that his head was killing him and that he needed to go to the doctor. When I walked over to him, he started screaming, "Don't you touch me, motherfucker! Don't even touch me!"

I knelt down to where he was sitting on the sidewalk and said, "The police want me check you out. Just shut up and let me have a look at you."

"You ain't touching me, damn it! You're the reason my head is killing me."

"Yeah, and according to your wife you're the reason your son is on the way to the hospital. So shut the fuck up and let me take a look at your head!"

I checked him out and didn't see anything wrong as far as I could tell. I told Charlie that I couldn't see any problems, wrong but he should probably take him to the hospital and let the doctors check him out. Charlie said, "I'll take him. But I have to ask what the hell you were thinking when you hit him in the head with that bottle?"

"I wasn't going to let him hit me. I just decided to give him a little O2 therapy."

Charlie laughed and said there wouldn't be any problems with it. I took the gear back to the truck and called us back into service. We headed back to the station and thought we would attempt to eat lunch again. This time we were able to eat without interruption. Once in awhile we do get to eat on time. Today wasn't the day, but sometimes it works out.

The afternoon was fairly uneventful. We slept through a boring video and then I went out to do my run. I try to run for thirty to forty-five minutes every afternoon that I'm on duty. Sometimes I am able to get my run in and sometimes I can't do it due to work. I finished and took a quick shower to cool off and so that I would not offend anyone. By then, it was time for dinner.

Bullet was cooking tonight and I was glad to hear that. Bullet's a hell of a cook and I look forward to his meals. I helped out a little by making the salad. I'm not the best cook in the world when it comes to cooking on a stove, so when I do any cooking at the station I usually do something on the grill. They don't seem to mind a steak or burgers from the grill. It's just a good thing that the weather stays fairly mild around here so I don't have to freeze my ass off when I do cook in the winter.

We had another great meal courtesy of Bullet. But I received no credit for the terrific salad. Once we cleaned up the mess we headed to watch some television. I figured it would be a pretty decent evening since there was a football game on. I can't help it, I love football. Let me put it this way; I even watch arena football so I must have a love for the game. We channel surfed until the game came on. When it started, we saw the opening kickoff and the tones went off. Figures. When there's something good to watch, eat or do, the tones always go off.

We were dispatched to a CPR in progress. When we arrived, we found no one performing CPR. The victim was a thirty-nine year old female. Her husband was running around screaming and getting in our way. I did my best to be tactful when I told him to move, but had no luck. Bullet couldn't even get him to stay out of the room. When a person goes into cardiac arrest, we have defibrillator pads that need to be placed in specific locations on the victim's chest. We have to remove the covering their chest to do so. He went nuts when he saw that we had ripped his wife's shirt and bra open. He must have thought he was dealing with a bunch of perverts but Bullet was able to convince him that we were doing what we had to do.

We shocked the woman twice. Then the defibrillator indicated no shock advised. We continued CPR. EMS arrived on the scene and hooked up their equipment. They started administering drugs into her veins. We helped in every way we could but the crew looked like they had everything under control. Bullet came in and told me that a police officer was keeping the husband out of our way. One of the paramedics said that she was asystolic, which in simple terms means that she had no pulse and no electrical heart activity. She was flat line.

I asked one of the paramedics what he needed from the truck. He said the stretcher, a backboard and the thumper. The thumper is a device that is hooked onto the stretcher and an arm is placed over the patient's chest. It has a device that comes straight down onto the patient's sternum and moves up and down. It can perform CPR on the patient, freeing the paramedics to do other things they need to do.

Bullet and I went out to the truck to get everything. I grabbed the thumper from the side compartment while Bullet got the backboard. A lady came to the truck and said, "I'm a nurse. Is there anything I can do to help?"

I told her, "Well, we have a thirty-nine year old female who is asystolic and unresponsive to any treatment."

She said, "What does asystolic mean?"

I handed her the thumper and said, "Here, you carry this."

I hate when people who should know what they are doing come up to help and have no idea what's going on. This woman was a nurse in a doctor's office and had no idea what asystolic meant. What made it even worse was when we found out she was a nurse in a cardiologist's office. If anyone should know what asystolic means, it is a nurse in a heart doctor's office. I decided right then and there that if I ever needed a cardiologist, I would remember that office and request anyone but that one.

We hooked the thumper up and packaged the woman for transport. She still was not responding to treatment. Both paramedics needed to be in the back with the patient so I drove the ambulance to the hospital. The husband wanted to ride in to the hospital with his wife and I had to fight with him to make him understand that he could not ride in the back. I finally got him in the front and made him buckle up. I got in behind the wheel and told the guys in the back to let me know when they were ready to go. A couple of minutes later they said, "Hit it!   10-18 traffic!" 10-18 traffic means lights and siren or emergency traffic.

We got to the hospital and the crew in the back jumped out with the patient. I took the husband to the waiting room and helped him get settled in. I warned the nurse about how hard he was to handle. She might want to get security up there to help out. I then wished him my best and headed back to the ER. I stopped off to see if the paramedics needed me anymore. When they said no, I went outside to wait on Bullet to come by and pick me up.

We headed back to the station and I started on the paperwork. After you use the defibrillator on a patient, there is a ton of paperwork. I had to do the state report, the fire department report and our version of the EMS report. Then I had to hook the defibrillator up to the computer and print out all of the readings that we had to take on the scene. Once that was done, I had to fill in the logbook. When I first became a captain, there wasn't nearly as much paper work as there is today. But everybody and his brother are so eager to file a lawsuit that you have to cover your ass in case anything happens.

By the time I finished all of the paperwork, the game was already late in the third quarter. I got a drink and kicked back in a recliner to watch the game. It was a pretty good game and I hate that I had to miss a big part of it. I was just getting settled in when the phone rang. It was Chief Williams wanting to know what I had planned for tomorrow morning. I told him that I was planning to work my part-time job but I could change that if necessary. He said, "We need somebody to go over to the town homes and help the TV station do a demonstration on smoke detectors."

I said, "I can do that. But I'll need a crew."

"Ask around there and see if anybody wants to make some over-time money."

I asked the crew on my truck and they said they would work. I told Chief Williams, "I have two others. Now what about a truck?"

"Go get a reserve truck from the city lot and be there at 9:00 tomorrow."

I wasn't sure what the program would be with the TV station but I figured I could stand to make time-and-a-half as opposed to making half that much working part-time. The town homes we had to go to were some old section 8 homes that were going to be demolished this coming winter to make way for more section 8 housing and, of all things, a golf course. People that live in section 8 housing don't ordinarily play golf. But that was the plan for the property. The fire department was able to get them for burn training and we have been using them for about six months. It's such a great opportunity that I can't believe the city would let us do it. Logic has never really played a part in this department in the past; it's just a little surprising that they would use it now.

I went to bed after the game. I wanted to be alert the next day since I would be taking inexperienced people into a burning room. I wasn't too thrilled about that, but it was their asses and I would have my air-pack on so I personally didn't have anything to worry about. The next morning the wake-up tones went off and I felt pretty decent. I couldn't believe that we didn't have a ton of rides that night but we lucked out.

I called over to he city lot and arranged for the truck to be delivered to us at the town homes. We went to get some breakfast and headed over to the town homes. When we arrived, the truck was sitting there with a note on the steering wheel to call when we were done and he would come pick it up. We waited around for a few minutes and the TV crew showed up. The reporter came up and introduced herself, "Hi, I'm Julie Marks and I'll be doing the reporting. This is Bill Roberts, my camera man, and Janine Williams, the producer."

I shook their hands and said, "I'm Dalton James, this is my driver, Bullet, and this is Firefighter Rick Johnson. So what do you need us to do?"

"Well, we came here yesterday and put up some smoke detectors in different locations and now we want to see which goes off first. All we need you to do is set a couch on fire and let us shoot video until they all go off."

"Simple enough. Bullet, go ahead and hook into that hydrant across the street and we'll take a line in."

We got everything set up and put our air-packs on. I had Bullet charge the line and we set the couch on fire. Rick and I were in air-packs but the reporter and her crew was standing there with nothing. I felt sorry for them but I wasn't about to give up my air-pack. I figured three victims would be enough.

It took a little while for the first detector go off. We let the fire go for a while and after ten minutes we were up to three detectors. The fourth and final detector still hadn't gone off and apparently they needed it for their story. After fifteen minutes more, the people from the TV crew were hacking and coughing and couldn't stand it anymore. The reporter told me that they wouldn't wait for the fourth detector and that I should just go ahead and put the fire out. I had been hitting the fire rolling on the ceiling to make sure the fire didn't get out of hand so it didn't take long to put it out. Once it was out, Rick and I went outside to take our air-packs off. The reporter was on all fours and puking her guts out. The reporter and producer weren't doing much better.

Bullet came over with the oxygen bottle and put it on the reporter. The cameraman took a few hits, as did the producer. They were starting to feel a little better and the reporter finally quit puking. That is, until Bullet came over eating a doughnut. When she saw that she hurled again. I had to step away; I didn't want her to hear me laughing. Bullet walked over to me with a big shit-eating grin on his face and raspberry jelly on his chin. I told him, "Damn that was cold. You ought to be ashamed of yourself."

He smiled and said, "I just thought she looked a little chunky and needed to lose a little more weight."

"Well after that last time she's definitely doesn't have any more chunks to blow. Let's go ahead and break this down. Oh, and wipe your chin."

Bullet unhooked the hose from the hydrant and loaded it back on the truck. Rick and I broke the hose from the truck and left it on the side of the street. The hose was already there when we arrived so we used that instead of getting the hose on the truck dirty. As we were finishing up, the reporter came over to me and said, "I couldn't breathe in there. How do you guys stand it?"

I reached down, picked up my air-pack and said, "You just have to have the right equipment. That's why we wear these."

She gave me a look that told me she thought I was a smart-ass. But I didn't care. I still had my breakfast and I didn't have to breath from an oxygen bottle when I came out. I called the city yard and let them know we were done with the truck. I told them that they would need to take it by headquarters so the air-pack bottles could be refilled and when they got there they would need to replace the oxygen bottle. They said that they would take care of everything.

I went home to take a shower. I called Rick Jackson at my part-time job to see what he had lined up for me. He said there wasn't much to do today so if I wanted to take the day off I was more than welcomed to. I didn't need my arm twisted to know what I should do. I went out and mowed the yard and took care of some other things that needed to be done at home. Then I picked the kids up and bought dinner on the way home.

As we ate dinner, I watched the news for the report on the smoke detectors. It was a pretty good story and the reporter covered up the fact that the fourth detector didn't go off. The one thing they didn't show was the reporter puking all over the place when she came out of the building. I guess that wasn't as newsworthy as it was funny to us. Some people have no sense of humor. Of course, Teri laughed about it when I told her what happened. She knows the reporter that did the story and doesn't like her.

After going to bed that night, Teri and I cuddled up with each other and the dogs and fell fast asleep. I always sleep better at home than I do at work. I think it's just being with Teri that makes me feel good and allows me to sleep better. It could have something to do with the fact that I worry about Teri and the kids when I'm at work. It's such a crazy world these days and I feel better when I can be there to protect them. I don't know if I could ever stop a burglar or not, but I know I would give my life for my family. I would do anything to keep them from harm.

    The next day I went in to the part-time job and redid a couch. I don't get to work with wood as much as I would like to there. I would rather refinish a piece of wooden furniture than reupholster a couch but you do what is assigned to you. I stripped the old upholstery off and that was tacky enough. But the new upholstery was even worse. I don't know why, but it seems that the more money a person has, the worse their taste in furniture gets. I don't care what anyone says, nobody needs a couch with huge orange and green flowers all over it. Tacky as hell.

    I did the usual routine on the way home. The kids and dinner and TV that night. There wasn't much on so Joey and I went to rent a movie. We watched the movie and then sent the kids to bed. It's always a struggle to get them to go to bed. I had to wake Nikki up to get her to go to bed. As she was stumbling down the hallway she kept complaining about having to go to bed when she wasn't tired.

    Once the kids were tucked in, Teri asked if I was feeling lucky tonight. I told her that I was feeling lucky if she was. That night we both got lucky. And there's nothing like an audience when you try to be with your wife. You have three dogs staring at you and they have looks on their faces like you're doing it wrong. Usually the two youngest dogs start wrestling with each other while Teri and I are trying to do something. I guess I shouldn't complain too much, though. It keeps them occupied and they don't bother us.

    Later, we let the dogs on the bed and we all fell asleep in a pile. Around 2:30 that morning I woke up from a dream that I was being choked. When I woke up my dog, Mojo, was asleep by my head with his front leg stretched out across my throat. I had to wake him up and make him move so I could get back to sleep. Teri woke up and wanted to know if everything was okay. I told her about Mojo. She said that I would be all right and held me close. Mojo didn't give a shit. He went back to sleep and snored to his hearts content. He didn't try to bump me off any more that night.

Chapter Sixteen

We had a class this morning. They had Station 15, Station 18, Station 21 and Station 26 attend a class on school bus fires. We had a one-hour classroom session then went out for a couple of hours to do an actual school bus fire. We decked out in our turnout gear and air-packs and got everything ready to go in. Then the firefighter handling the class, Josh Mitchell went in to set the bus on fire.

Josh is a very large man bordering on the obese and when he set the fire he had some difficulty getting out of the bus. Josh cannot get the buckles on his air-pack hooked together. When he wears it he just pulls the shoulder straps tight and it stays on. When Josh set the fire in the back of the bus, he backed out and got one of his straps hung on the door handle of the bus. He started out the door but he could not move. He had both arms and both legs moving and had smoke billowing out of the bus over his head but he could not move. Two guys from Engine 26 ran up to help him get out. When we first saw him, I thought he was wedged in the door. The rest of us were laughing too hard to be of any help.

The guys from Engine 26 finally got the strap off of the door handle and Josh fell face first to the ground. When he rolled over on his back he was kicking his legs and waiving his arms, trying to get up. I looked at Bullet and said, "I'll be damned if he doesn't look like a tick about ready to pop."

Bullet went over to Josh and stared down at him for a moment. Then he looked at me and said, "Damn, Cap, you're right. He does look like a tick."

Josh was pissed. He doesn't like to be picked on about his weight. I don't know if he thinks that no one notices his size or not but he doesn't have much success trying to hide it. When you are six feet, two inches tall and weigh 350 pounds, it's going to show.

I was stationed with Josh about five years ago. He always slept in his briefs and when he would get up in the morning and walk down the hall to the locker room, you couldn't tell if he was wearing anything or not. His gut hangs down to his thighs and it's all he can do to get it lifted to buckle his pants.

We went in and put the fire out. It didn't take much to put it out and overhaul the bus. When we were finished, we noticed that Josh left to go to the training building. That's when we went to work on his turnout gear. I went to the truck and got some duct tape and a magic marker. The back of our turnout coats have our names written on them, so I covered his name up with duct tape and wrote "Tick" on it. Then I took the magic marker and wrote "Tick" on his helmet. When Josh came out of the back, he grabbed up his gear and didn't notice what we had done. He had his crew get on the truck and headed back to their station.

We finished cleaning everything up and headed back to the house. None of us felt like cooking lunch so the squad stopped and picked up some burgers. While we were eating lunch, we started talking about some training accidents that we've had in the past. We've had the usual types of accidents. People falling down, hoses busting, things like that. But the best one that I remembered was the time the training chief nearly blew himself up.

We were burning a house on the south end of town one morning. There were four people trying to run the class and one didn't know what the other was doing. It's like the old saying; we had too many cooks in the kitchen. They were arguing back and forth about how the class should be run. We had four different versions of what we were supposed to do. After half an hour of this, one instructor went in and poured diesel fuel on some straw in the kitchen. When he came out, the instructors got into an argument again about what to do during the class.

This argument went on for another forty-five minutes while the fumes in the house were building up. After they finally figured out what they wanted to do, Training Chief, Steve Williams, went into the house to light the fire. Before doing so, he poured more diesel fuel on the straw, not knowing that it had already been taken care of. He couldn't smell it because he had an air-pack on. He then went towards the side door of the kitchen and struck a match to light the fire. Just as quick as he lit the match, the fumes in the room went off and the kitchen exploded in flames. The force of the explosion shot Steve straight out the door and into the side yard. The entire house went up and all we could do is surround the house and let it burn to the ground. That was the original reason we were using the house, to burn it to the ground. But we wanted to use it for a while to train with.

Steve turned out to be okay. Just a little sore from landing in the yard and bouncing about three times. We let the house burn to the ground while we protected the exposures. I was told to take a nozzle and work it back and forth on the mortar in the chimney. With enough pressure behind the stream, you can remove the mortar and take the chimney to the ground.

I had worked the night before and had four rides. The only reason I was there was because they needed some experienced firefighters to take a group of rookies through the house. I was dead tired and while I was working the nozzle on the chimney, I noticed something out of the corner of my eye. When you are tired, your body tends to follow your eyes. When I looked over at what I saw, my hands followed my eyes and I nailed a battalion chief in the back with the stream. What caught my eye was the white shirt he was wearing. He was on the backside of the house and when I looked at him I nailed him full stream.

He was a little pissed off about it but when I asked where his turnout gear was, he quickly made up an excuse and left.

That afternoon, we cleaned the station and truck. I worked on the training log and took care of some other paperwork that needed to be done. I had to do an evaluation on Rick so he could get a raise. It wouldn't be much of a raise, but I guess a few cents is better than nothing at all. The city has gotten so tight with raises that we don't get very much these days. They tell us they cannot afford to give us larger raises, yet the city council can afford to go to a golf resort for "special meetings." They cannot afford to give us decent raises, yet they can afford to give tax incentives to companies coming into town.

The tax incentives wouldn't piss me off so much if the people that worked in these companies actually came from Mannington. But they don't. Only two percent of the people that work in these companies come from Mannington. The rest of the people come from surrounding towns and do not pay taxes here. What the city council doesn't think about is the future. What do they think will happen after the twenty-year tax incentive is over? These companies won't stay here. They'll move to the next city that gives them tax incentives. Once that happens, there will be a lot of abandoned buildings on the north end of town and it will look like the downtown area. Empty, dilapidated buildings and no jobs in the area. The city council needs to look at the future. There is no obligation for these companies to stay once the tax incentives are over. If the city wants to give tax incentives, they need to get some kind of guarantee that the companies will stay in the area once the incentives are over.

We were sitting around waiting on dinner when police officer Paul Burrows stopped by to get a cup of coffee and shoot the breeze. Jack is a canine cop and has a great dog named Nick. Nick is a Belgian Shepherd and is one hell of a good police dog. I've seen Nick in action and I can tell you, he is one hell of a fighter. I've seen him control two suspects at the same time.

We started talking about how the dogs sniff out drugs and Mike Berrier said he didn't think the dogs were smart enough to find anything. He pointed out that Nick didn't look like he could find his own ass with both paws, much less drugs. Paul decided to put on a little demonstration for us.

Paul took Nick to his car and came back with a big Tupperware container full of drugs. He explained that the drugs had been seized a long time ago and he was allowed to use them for training his dog. Of course, there is an accountability system in place for these drugs, so he can't use them. He pulled out a small bag of marijuana and hid it behind a bookcase in the kitchen. He then went back to the car and brought Nick back with him.

Paul gave Nick the search command and Nick started walking around the room in a zigzag pattern. He went from one side of the room to the other and kept searching until he found the pot. The whole time, Mike was saying how it was a game. That Nick knew where the drugs were all of the time because that is where Paul would hide the drugs every time. While he was talking, I found a small bag of cocaine in the container and slipped it into Mike's back pocket.

As they were leaving, Paul packed up the pot in the container and told Nick that they had to go. Nick started for the door when he caught a whiff of coke coming from Mike's back pocket. Nick went nuts and pinned Mike against the counter. Nick is trained to start clawing and biting at the point of drugs. When he started biting Mike's ass, we all fell out from laughter. Paul searched Mike and found the bag of coke and had a very serious look on his face. I don't know if he was ready to arrest Mike or not, but we couldn't tell him what we did because we were laughing too hard.

Finally, I was able to compose myself long enough to tell Paul what I did. He thought it was funny as hell, but he did count all of the drugs before he left. The best part was when Nick walked over to Mike and sat down in front of him. Mike looked at him and asked, "What the hell are you looking at?" Nick had an apologetic look on his face and held out his paw, ready to shake hands. When Mike shook Nick's paw, Nick gave him a kiss on his cheek and headed to the door. It was as if Nick was saying, "No hard feelings."

The tones went off as we were finishing dinner. We were dispatched to a reported house fire on Main Street. When we arrived, we found the front of the house was rolling pretty good. Rick caught the hydrant. I put my airpack on and Rick and I took the line to the door. Engine 21 arrived and I had them act as the rapid intervention team. As we were going in, Chief Williams arrived and took command.

When Rick and I entered the room, I popped the nozzle on the ceiling to knock the fire back. The room wasn't fully involved yet so we were able to get in there fairly easily. We worked our way back to a corner and found the seat of the fire. We knocked the fire out in a matter of minutes. There didn't appear to be any structural damage to the house, but the fire had been burning against the ceiling for a while. I called for a plaster hook and started pulling the ceiling down. I worked my way across the room, pulling what was burned to the floor. After getting halfway across the room, I jammed the hook into the ceiling near the attic access and pulled down the plaster. There was something else in there with the plaster. I took my flashlight out and looked around on the floor until I found what fell. It was a dildo. This thing was huge. It was two feet long and six inches in diameter. I couldn't believe that any woman would want to use something that big on herself but there it was, lying in the rubble on the floor.

As I was pulling the ceiling, other firefighters were in and out with pails and shovels, taking the debris out of the house. We had already cleared the room of furniture and anything that could be salvaged. I needed to move the dildo because I didn't want it to be carried out into the front yard with the rest of the debris. That would embarrass the woman that lived there and I didn't want that. The problem is, I really didn't want to touch it either. I took my rubber gloves out of my pocket and picked it up. I put it on the mantle across the room and left it there for the homeowner. We pulled the rest of the ceiling and didn't find any more surprises.

We took care of all the debris and didn't find any fire extension beyond the room. We broke down our hoses and loaded back everything that we could. We then headed back to the house to load the truck and clean the hose. The problem is, before I left I forgot to tell the woman what I found. I started laughing on the way back to the station. Bullet looked at me and asked, "What's so funny?"

I said, "When I was pulling the ceiling in the house, a huge dildo fell out of the ceiling. It was two feet long and six inches wide."

"Damn! But what are you laughing about?"

"Well, we didn't carry it out into the yard with the debris. I took it and put it on the mantle and meant to tell the owner about it. I would like to see the look on her face when she sees that dildo standing tall and proud on the mantle of her living room."

Bullet laughed and said, "Just think about what will happen if she doesn't see it. The insurance man comes in to look at the damage and then sees that thing sitting there."

"Maybe she could call it modern art or something."

We loaded the truck and took care of the dirty hose. Then I went in and did the report on the fire. The only thing I didn't put in the fire report was the dildo. There is a section to put unusual objects found in the fire, and I was tempted. But as a courtesy to the homeowner, I didn't.

By the time I finished my shower, it was 11:30. I decided to head to bed to try to get some sleep. I was hoping for a slow night because I was beat. House fires are a lot of work and they can wear a body out. Especially when you have to pull a ceiling. I didn't really want to work that much, but the department is in a manpower shortage so we all have to do more work on the fire ground. That's just part of the job, I guess.

The next morning we did our shift change and I headed to my part-time job. I only had a couple of things to do and I was able to take my time. It was a slow day but I was able to drag the jobs out long enough to get a full day in. I hate doing that but when I hurry through a job, Rick Jackson will come along and tell me to slow down so I can get a full day's pay. Of course, it never works when I tell him that he could just go ahead and pay me whether I work or not. But I can't complain too much. I don't have to do much and I can have time off whenever I need it. I guess it's not such a bad situation.

I picked the kids up after getting off work and Nikki needed some things for school. I took her to the store and we bought the supplies that she needed for a project. We then surprised Teri at work and went out to dinner. We met her in the parking lot at the hospital and took off to get pizza. It was Friday and neither of us had to work the next day. The kids didn't have anything going on for Saturday so we stayed out pretty late.

When we got home and the kids were off to bed, I told Teri about the fire the night before. She was as shocked about the size of the dildo as I was and didn't believe me at first. But I convinced her about the length and girth and all she could say was ouch. I had to admit that it would be pretty painful to use that size on yourself. But, then again, it could have been a gag gift or something that she just had as a joke. Maybe she didn't use it at all. But then again, you never know about some people.

The next day, Joey and I worked in the shop for a while. He wanted some shelves for his room so we went and bought the lumber and then to the shop for a little bit of woodworking. Joey loves to work with wood, he just gets a little impatient when he has to let the glue dry. But I was able to occupy that time with some carving, so he didn't get bored. Joey has a very short attention span and doesn't like to concentrate on things for very long. But you give him a knife and a piece of wood and he is content for hours. I'm not sure what it is about woodworking that he loves so much, but he is definitely hooked. But, I'm hooked on it too, so maybe that explains it.

We did steaks on the grill for dinner and ate outside on the deck. Joey wanted to throw the football after dinner so we went in the front yard. We would have done it in the back yard but the problem is that we have to dodge so many "land mines" from the dogs that we usually go to the front yard. Makes for a better time than having to go scrape your shoes every five minutes.

We had fun throwing the ball and I had a blast watching him try to kick the ball. He teed the ball up and walked back about fifteen yards. He got a full head of steam up and ran like a bat out of hell towards the ball. He attempted to kick the ball and missed it completely. He ended up flat on his back. After finding out that he was okay, I laughed my ass off. I couldn't help it. I think it might have pissed Joey off a little but we joked about it for a few minutes and he felt better.

That night we went to bed and Teri and I had a scrimmage of our own. One thing is for sure, neither one of us dropped back to punt. We then fell asleep in each other's arms and slept the night away. Around 2:00 in the morning, our dog, Jake, decided he was a little lonely and wedged his way between us. I guess that he had had enough of the mushy stuff and decided to break it up. We both just rolled over and made room for him. The next morning I got up and went in to work.

Chapter Seventeen

Sunday mornings aren't too bad around the station. Of course, I would have rather have been at home with my family, but when you get to eat a firehouse breakfast it makes it a little better. Bullet made eggs, bacon, sausage, pancakes, hash browns, biscuits, gravy, and grits. By the time you eat all that, you're ready to go back to bed to sleep it off. It was good eating but I when I finished, I knew what a stuffed turkey feels like on Thanksgiving.

I went into the classroom and kicked back in one of the recliners to "watch a little television." TV sucks on Sunday mornings. So once the breakfast kicked in, I didn't have to worry about it. What I ended up watching was the insides of my eyelids. That's not a bad way to spend the morning. It makes the time pass a little quicker. The only time I woke up was when Bullet punched me in the arm for snoring. He couldn't hear his fishing show. I told him my snoring was probably more interesting than any fishing show could be. I like to go fishing and I like to play golf, but to me there is nothing more boring than sitting in front of a television and watching it.

When lunchtime rolled around, I was still too full from breakfast to even think about eating. I stayed in the classroom and watched the football pre-game shows. It's not a bad deal when you can sleep all morning and watch football all afternoon. Then you eat dinner and watch another ball game that evening. I wish that's all we had to do that day, but of course the tones went off around 5:30 and we had to go out on a call.

We were dispatched to an unknown rescue call at a nearby residence. When we arrived, a man was standing on the front porch laughing and talking to a couple of neighbors. I walked up to them and asked, "What seems to be the problem?"

The man said, "You might as well come with me. You wouldn't believe me if I told you."

We walked in the house and to the bathroom. There was a woman sitting on the toilet and she did not look happy at all. The man wouldn't enter the bathroom, but told me in the hallway what happened. He said, "We're in the process of remodeling this house and I was working on the bathroom. The toilet seat was cracked so I wanted to fix it. She was going to her mother's house for the week and wasn't supposed to be here. I used some epoxy paint on the toilet seat to fix the crack. It stunk the house up so bad that I had to leave. After painting it, I left and went to Kelley's Bar to watch the ball game. When I came back, I found her here. She had a fight with her mother and after the drive back, she had to go to the bathroom. I didn't leave a note in there about the paint because she wasn't supposed to be here."

I asked, "And she can't move can she."

"She's stuck bigger than shit. No pun intended."

I went in and found that the man had put a bathrobe on his wife. I knew she would not want us to pull her loose from the seat. It would rip the hide right off her. I wasn't about to go that route. She asked me, "Don't you have chemicals to melt this thing?"

I said, "Well, we can melt it easy enough, the only problem is it would burn you, too. We'll have to cut the seat loose and take you to the hospital."

"I can't go out of here with this thing on my ass. Isn't there some other way?"

"I'm afraid that we don't have the right stuff here to dissolve the glue."

"Can't you get it? You have radios."

"Ma'am, this will have to be removed by a doctor. I can't do it here."

"Then why the hell did we call you?"

"Because the doctor doesn't make house calls."

I went to the truck and got the bolt cutters and a screwdriver. I had to cut the lid loose to get to the screws in the back. The whole time she was warning me that I had better not be looking at her ass. I assured her that all I was doing was concentrating on the screws in back of the toilet seat. Besides, my eyes couldn't take in a view of an ass that big. I had no interest in seeing her big ass at all. All I wanted to do was get her loose from the toilet and get back to the station to watch the game.

Once we had her loose, we gave her some privacy to clean up and get ready for the trip. EMS had arrived and were waiting outside on the porch. She was pissed when she found out that we weren't bringing the stretcher into the house. The hallway was too crowded to get the stretcher in there with all of the junk from the remodeling. Besides, she was ambulatory and could walk to the truck. We put a hand on each arm and helped her walk to the truck. She wasn't happy about that but she did it with all of the dignity that she could muster.

When she was loaded on the truck, her husband came up to me and asked, "How long do you think it will take to get that thing off of her?"

I said, "I don't know. Why?"

"Well, I was just thinking that if it was going to take a long time I could catch the rest of the game."

"Sir, can I give you piece of advice?"

"Sure."

"If I were you, and my wife was as pissed at me as your wife is at you, I would get my ass over to the hospital as quickly as I possibly could. There'll be other games."

"Yeah, you're probably right. Maybe I better go with her."

"Well, before you leave, you might want to take some clothes with you."

"Why?"

"Well she probably won't appreciate having to come home in her bathrobe."

"Oh. I'll get some just as soon as I start the VCR to record the rest of the game."

As we were leaving, I said to Bullet, "Man, if I was that bad about football, I don't think I could live with myself." Don't get me wrong. I love watching football. It's my favorite sport, but I can walk away from it if necessary. I don't let it dictate how I live my life and if there is something that I want or need to do with my family, I can miss a game or two. Sports will never dictate how I live my life.

We got back to the house and I did my report. I couldn't wait for them to read it downtown. For some reason, this station always seems to get the strange calls. I think the best one was a teenager who got a little too friendly with a vacuum cleaner. He just didn't know what would happen when he turned the vacuum into his date. I guess instead of having a date with "Rosy and her five friends" he had a date with "Miss Hoover." Well I can understand that. It did have extra suction power.

We arrived to find the mother and her son sitting in the living room. She was giving him hell when we walked in. I asked her what the problem was and she pointed to her son and said, "You tell him!"

He mumbled something about being stuck. When I asked him what he said, he told me, "I'm stuck."

I looked around him and asked, "Stuck to what?"

"This vacuum cleaner." He pulled the blanket back and showed me that he had his penis stuck in the nozzle of the vacuum cleaner. I couldn't help but laugh. The boy didn't like me laughing but I couldn't help it.

I gave Rick the assignment of looking after the patient. When I told him, he said, "Thanks a lot, Cap."

I just smiled as he started assessing the patient. I got some information from the mother about her son and the vacuum cleaner. It seems this particular brand has an anti-clog device that is built into the nozzle. It is a flat piece of metal half the diameter of the nozzle opening and it is made to spin and break up any clog that might form. I also found out that when the boy started the vacuum, he already had his penis in the nozzle. Apparently the metal turned and jammed on his penis and now he was stuck and starting to swell.

Rick turned to me and without thinking said, "Cap, this things going to have to be cut off." The boy almost fainted when he heard that and started crying. Rick turned to him and said, "No, no, no. Not your penis. The hose." The boy calmed down and was able to relax a little.

We cut the hose off the nozzle and had to figure out a way to cover him up. His mother said that she was not going to pay for an ambulance. She would drive him to the hospital herself. The only thing he could wear was her bathrobe. He doesn't have one and there was no way to get a pair of pants zipped over the nozzle. He had to walk from the house to the car with her frilly bathrobe on and that big nozzle sticking out the front. I couldn't resist the temptation to see the emergency room staff watch this kid walk in with his little friend.

We got on the truck and took a shortcut to the hospital. We waited in the admitting room to watch the action. The nurses kept asking us what we were doing there and all I would say is, "Just wait." When the boy and his mother arrived, she walked him into the waiting room in front of the staff and everyone sitting there. As he passed, everyone who saw him started laughing. They may not have known the story behind the nozzle, but they all knew what was stuck in there. Everyone was rolling with laughter as he walked by. I was laughing, too, when they walked in and the nurses saw him.

The nurses took him into an examining room to check the situation out and his mother came up to me, and said, "I can't believe he would do such a thing."

I told her, "Just look on the bright side. At least you don't live on a dairy farm." She looked confused and all I said as I walked away, "Milking machines."

I received a ton of phone calls about that call and I'm sure I'll receive quite a few because of the woman and her toilet seat. I received a lot of phone calls about another ride we had involving a child with his head caught in the banister at his home. At least, that's how the call came in.

When we walked into the house to rescue the boy, we saw him standing in the foyer of the house with peanut butter on his ears and crying. I asked him if he was the one who was stuck in the banister and he nodded his head. I figured there was nothing to do there so I just needed to get some information from his mother and we would be on our way. I asked him where his mother was and all he could do was point. When I saw her, I could only laugh.

She now had her head caught in the banister. As I choked back the laughter, I asked her if she was okay. She said, "No, I'm not okay. My heads caught in this banister and I can't move."

I asked, "Are you breathing okay? Are you in pain?"

"I'm breathing okay and I'm not in any pain. I just need to get out of here. Please get me out of here."

As we were getting her out, I said, "The call came in as a child being stuck in the banister. How did you get in there?"

"He was stuck in here. He was screaming and crying and I had to get him out. He didn't want to wait on the fire department to come. He just wanted out."

"So that's why you put peanut butter on his ears?"

"Yeah. And once I put the peanut butter on his ears, I pushed his head out."

"But how did you get your head stuck in there?"

"Well, when I pushed his head through, I lost my balance and fell into the banister. My head went through and now here I am."

I was able to hold my composure while we got her out. We were able to remove one piece of the banister and that allowed her to get her head out. Once she was out, we put the piece back in and you couldn't tell it had been removed. I quickly got the information for my report and practically ran back to the truck. People don't like to be laughed at so I did my best and was able to hold the laughter back until I got in the truck. As we were leaving, Bullet said she was on the porch but I couldn't help it. I was laughing so hard that I knew that she heard me. She never called to complain, but I'm sure she heard me. Sometimes you can't help it. Sometimes you just have to laugh. I'm sure that she didn't find it funny then, but hopefully by now she can look back on that day and at least smile about it.

I finished the report and went to the kitchen for dinner. We talked about the call and laughed about it. The ladder captain, Bill Waddell, would have loved to have been there. He loves big women and he would have loved her big ass. The best day of Bills life was the day we hired a stripper to meet him at his house and dance for his birthday. This woman weighed over 300 pounds and we knew that was what Bill liked. I'm not sure why, but he likes his women big. It doesn't match very well, though. Bills six feet, two inches tall but he only weighs 165 pounds. When he and his current girlfriend stand next to each other, they look like the number ten.

After dinner, I went to watch the evening game. Jack Walters, from EMS came by to see us. He introduced us to his new partner, James Stokes, and we shot the breeze as we watched the game. James is new to EMS and has never worked in emergency services before. We told him some of the stories and what to expect from the job. Mike

Berrier walked in and sat down. James recognized him from the jujitsu class they were taking and they started talking about their classes. We knew that Mike was taking jujitsu. He had been bragging about the classes and how good he was doing. He would grab people and do some of the holds he had learned. We were all getting a little tired of it. He's too fat to eat Chinese food, much less do jujitsu. He does good to move at all unless it's on a fire call or dinnertime.

Mike was talking about a new wrist hold he had learned, but James just couldn't figure out how he did it. Then he asked if I would help demonstrate the move. I told Mike, "I'm willing but if I end up getting hurt, you're a dead man." I didn't really want to do it, but I figured this would be a good chance to teach him a lesson.

Mike took me by my right wrist and pulled it around me. It didn't hurt but I couldn't let it slide. When he let go of my wrist, I pulled it close to my chest and rubbed it. When I was in college, I broke my right wrist. The radius, the bone behind the thumb, was broken into three pieces. Ever since I had the cast taken off, my wrist pops whenever I want it to. I was still rubbing it and Mike looked at me and asked if I was okay. I held the arm out and started popping my wrist. I looked at it in "shock" and said, "Man, what the hell did you do to my wrist?"

Mike looked like he was about to pass out when he heard the popping. He just knew his career was over and that he would be facing a lawsuit. I looked at Jack and winked. He came up to me and helped me walk out of the room. James was starting to follow when Jack turned and said, "You just stay here. I think you two have done enough to this poor man."

We walked out to the engine room and I had Jack splint my wrist and put a sling on me. Bullet came in and saw Jack putting the sling on me and he asked, "What the hell happened to you?"

I said, "Nothing. Mike was demonstrating a new wrist hold so I'm teaching him a lesson. I'm not hurt. I just want him to think I am."

"You know, he's getting out of hand with all of this jujitsu shit. I'm about tired of it."

"Well, maybe this will show him that he needs to stop. At least he won't be doing any more demonstrations on me."

"He don't touch me too much."

"Well, what do you expect. He's scared of you."

"Why do you say that?"

"I told him how you got your nickname. I guess he figures that anybody that would shoot his own truck wouldn't think twice about popping a cap in his fat ass."

I went back into the classroom and Mike and James were sitting there. I sat down in a recliner and held my wrist. I threw in a few moans for good measure and Mike was getting very nervous. I told him to go get his captain and bring him back with him. While Mike was gone, Jack told James to go wait in the truck. Before he left, James wanted everyone in the room to know that he had nothing to do with it. That it was Mike's idea, not his. James just said, "Go sit in the fucking truck. I'll deal with you later."

Mike came back with Bill Waddell and I made Mike explain what happened. While he was explaining he must have told Bill that it was an accident ten or fifteen times. He also made sure that Bill knew that I willingly let him put the wrist hold on me. I told Bill, "Look, this shit has to stop. He's going to end up hurting someone one of these days."

Bill had Mike leave the room, then said, "Well, apparently he already has."

"Well, actually, I'm not hurt at all. I've been able to pop my wrist since I broke it back in college. I just want him to think I'm hurt. Maybe it'll teach him not to do this shit anymore."

"Tell you what. I'll take him in the office and chew him out a little. Ream him real good. Do a letter to the chief. The works."

"Okay, but don't tell him I'm not hurt. Just let him sweat for awhile."

"I'll do it. But what if you guys get a call."

"If it's a fire call, he'll know. If it's a med call, we'll tell him all I did was supervise."

Bill took Mike in the captain's office and closed the door. It didn't do any good to close it, however, because you could hear him screaming at Mike all over the station. He kept Mike in there for two hours while he chewed his ass out and wrote a three-page letter to the chief. We could see through the window that Mike was sweating it out pretty hard. Then Bill did something that I wouldn't even have thought of. Bill told Mike to start packing his locker up.

Mike walked to the locker room like he had just lost his best friend. Rick came to the engine room with a football and was throwing it around behind the trucks. I got in on the game of catch while Bill told us about Mike. He told me I really needed to let him know I wasn't hurt before he got in his car and left. I told him I wouldn't let it get that far as I missed a pass from Rick. The ball bounced off the wall and hit the bay door. It made a loud bang and finally Mike came out to see what was going on. As he came out the door, I was just raring back to throw the ball. I stopped in mid-throw and slowly looked over my shoulder at him. All I could say is, "Hey, Mike. What's up?"

Mike didn't take it very well. He was furious. I didn't realize that he had taken the shelves apart in his locker. He had taken everything apart in his locker and had it all packed up. He was ready to start moving boxes to his car when he came out. He was storming around threatening to kick the shit out of all of us. Bullet made some comment that only Mike heard and Mike grabbed him. Bullet said, "You know better than to grab me like that. I've got something in the truck that'll take care of your fat ass."

When Mike realized who he grabbed, he let go and jumped back. Mike was truly afraid of Bullet and Bullet liked that. He doesn't like to pick fight with anyone and we all know that. Bullet's philosophy is don't grab somebody unless you're ready to go the distance. Mike took off running from Bullet, figuring he was a dead man. Bullet never moved. He looked at me and said, "He ain't worth the effort."

We went back in and watched the rest of the game. We kicked back and let Mike cool off. He had plenty of time while he unpacked and redid his locker. Rick went in there a couple of times and kept coming back giving us reports on how Mike was throwing things into his locker and muttering to himself. I told Bill, "If he doesn't get it neat, we could have a locker inspection on Tuesday." Bill thought that would be a good idea.

After the game, I called Teri and told her about the call we had and what we did to Mike. She thought that was a cold-blooded thing to do to him. But she also knows how Mike. Before I went to bed I went and talked to Mike. I told him that he needed to cut the shit out before he really did hurt someone. It was okay if he did that stuff at class, but he needed to stop it around the station. I don't know if it will work or not but you never know. Miracles do happen.

We lucked out and didn't have any calls that night. I couldn't believe we got off with only one call for the shift. It usually doesn't work out that way. Ordinarily, if we have a quiet day, we end up running all night. But once in a while, St. Florian, the patron saint of firefighters, smiles on us and lets us have a quiet night. The next morning, shift change went smoothly and I left the station for my part-time job.

When I arrived, there was a sign on the door that said they were closed. I saw a note under the closed sign that said they would be closed for the week because there was nothing for us to do. I drove home and started wondering what was happening. Business had slowed down so much that they had to close for the week. I wasn't sure what would happen with the company but I didn't have too much to worry about. Teri and I really didn't need the money that badly. She and I both make decent money and our bills were caught up. Of course, if she wasn't head nurse at the ER, we would need it. I'm not ashamed to admit that Teri makes more money than I do and that's one reason we don't really need the money that bad. As a matter of fact, the only reason I was working part-time was to pay the house off early. We didn't have many more payments to make before it would be ours.

When I got home, I saw the answer machine blinking and checked the messages. It was Rick Jackson telling me that he forgot to call me the day before, but I was off for the rest of the week. He said he would let me know what was going on later in the week. I decided to take care of the yard work before it got too hot outside. It doesn't take long to mow the yard and I was done in less than two hours. After finishing, I stayed inside and enjoyed the air conditioning.

I went and picked the kids up from school. I didn't see any reason to have them go to the YMCA when they could do their homework here. They were a little surprised to see me when they got out of school. We went home and kicked back for awhile. I did burgers on the grill for dinner and they did their homework while I cooked. Teri came home and we had dinner on the deck. We sat around after dinner and watched Joey and Nikki throw the ball for the dogs out back. That lasted until Nikki stepped in one of the dogs' land mines.

The kids went to bed early that night and Teri and I watched a little TV before going to take a shower. We like to be earth conscious and we take showers together whenever we can in order to save water. If more people took showers together, there wouldn't be as many water shortages. Besides, there are just some places that are hard to reach. It's good to have people there in case you need help.

The next day, I slept in late and hung around the house doing odd jobs. I did go out to the shop and work on a few things out there. I have fun making things for people. The thing I like doing most is woodcarving. I don't know what it is about carving that attracts me so much, but I love doing it. And after ten years, I think I'm getting pretty good at it. I don't claim to be the best carver in the world, but I am getting better at it.

One of the things I like making the most is walking sticks. I must have fifty sitting around the house and my shop. That's not to say that I have never made them for other people. It's just that I sometimes have a hard time parting with them. I don't need a cane or walking stick to get around, but it's nice to know that I am covered when I get old enough to need one. The cane I started working on today is for a retired firefighter, Danny Steele.

Danny had to go out on disability after getting injured on the job. He was ventilating at a working house fire when the roof collapsed and took him with it. It should never have happened because he shouldn't have been up there. The problem is, there was a crew inside damn near burning up while looking for a small child. The roof collapsed and Danny landed on his left leg. Messed it up pretty bad. He broke both bones in the lower leg, the femur, the hip and blew his knee out as well. So far he has had seven operations on the leg but he will never regain full use of it.

I decided to pick the kids up from school again. Since I was working in the shop, I knew the kids would like to get out there and do something. Joey likes woodcarving almost as much as I do and Nikki enjoys painting and gluing pieces of wood together. When we got back to the house, the kids went to the shop and started working on their projects. Joey has been working on a cane of his own. He is getting better and better at carving. I can't believe some of the pieces he has turned out. One of these days I'll have to quit teaching him and let him teach me.

Nikki started working on a couple of pieces of wood that I cut out for her. She likes to make things for her grandmothers to put out in their gardens. And everything that Nikki makes for them proudly goes on display no matter what. And I have to admit that she is good. She knows what has to be done to protect her paintings and she makes sure it gets done. She never misses a step when it comes to finishing a project. The only problem is she gets a little impatient when it comes to drying time. That's why I always make sure she has three or four projects in the works so that she doesn't get bored.

We worked out there for a couple of hours, then I had to make them quit so they could go do their homework. There was a message on the answering machine from Teri. She told me she would be getting dinner on the way home so I didn't need to worry about making anything. I could live with that. I went to my keyboards and played some music while I waited on her to get home. Joey needed to use the computer to do some research for a paper he had to write. I got him started on his research and he worked on-line while I played the piano.

Teri came in with pizza and we ate dinner in front of the TV. Joey had already finished his homework. Nikki had to finish some math homework home dinner. That night, we watched wrestling on TV. We all know that wrestling is as fake as hell but we still enjoy it. I like watching it to see how many ex-professional football players are wrestling these days. I usually watch wrestling when I miss cartoons. I figure one is just as believable as the other is.

Once the kids were in bed, I got things ready for work the next day, and then took a shower before bed. Teri was already asleep when I crawled into bed beside her. Actually, I crawled between her and the dogs. I had to fight with Jake to get him to move out of the way so I could lie down. He gets hard headed when he's sleepy. I finally got him to move and I was able to go to bed. I cuddled up next to Teri and fell asleep.

Chapter Eighteen

The next day I went into work. Mike was sitting outside when I arrived and he didn't have a lot to say to me. He was still pissed about what we did to him Sunday. I just hope he remembers everything that he had to go through when he thought hurt my wrist. then, he might stop all the bullshit he's been pulling around the station. Mike is a good firefighter but the problem is, he is too much like a kid. He doesn't know when to quit playing. Now maybe this will help him focus on his career.

We had a morning class on natural and LP gas. Nothing new. Go in with two hose lines creating a water curtain and a man in the middle with tools so that he can shut the line down. It's not as easy as it sounds, though. If the water stream is lifted just a little bit off the ground, the fire can get underneath it and bite you. The fire can also come around the water curtain and get the guys on the hose lines. The fire burns above three thousand degrees and it will definitely cause severe damage to whomever it comes in contact with. The two hose streams need to be on the ground. There can be no gap in between them and they must form a wide curtain. Otherwise, it could be deadly for the people working on the leak.

Fortunately, we each have an extra set of turnout gear because what we wore to the class was soaked all the way through. It's bad enough to have to wear turnout gear in this heat, but when it is wet and heavy, it's even worse. I changed out all of my gear and entered the training class in the computer. I figured I would get the paperwork done early and get it out of the way. We didn't use any gear off our truck other than our turnouts and air-packs. When I finished everything in the office, I went to the kitchen for lunch.

We started a class that afternoon on ladders. Nothing new. We reviewed the process of carrying and raising ladders. We had only been in the class for half an hour when we were called out for a car fire. We only sent one engine because the car fire was more than ten feet from the house. When we arrived, I assessed the scene and saw that there was nothing around the vehicle.

When I got off the truck, Rick came off with his airpack on and went to the back to pull the attack line. We had a fourth man on the truck that day, Tony Jackson. Tony has been on this department for ten years and ended up getting a transfer to our station. Tony and Rick started working on the car fire. It was burning under the hood. Tony took the hose line and started hitting the front of the car. He told Rick to pull the hood release inside. Rick opened the door and smoke billowed out. The car was full of smoke and Rick couldn't see very well. Instead of pulling the hood release, Rick pulled the release for the parking brake.

I got on the radio and said, "Engine 18 to fire communications. Engine 18 is on the scene. We have a rolling car fire."

What I meant was the fire was rolling. When I turned around, the car started rolling down hill. I took off running down the hill as fast as I could. I ran the one hundred-yard distance to the "T" intersection and ran out in the middle of the road to stop traffic. I stopped the only car that was coming down the hill. The driver looked pissed that I stopped him and was yelling at me for it. As he was yelling, the car rolled across the road behind me and hit the curb so perfect that it missed the telephone pole on one side and the bench at the bus stop on the other side. It ended up in a park at the bottom of the hill.

The driver of the car smiled and gave me the okay sign. Then he started to move but I stopped him again. He looked at me in confusion and threw his hands in the air.

He started yelling again when the engine pulled out into the intersection with the attack line dragging behind it. When Rick and Tony got to the truck, I motioned for the driver to go ahead and drive on through. At this point, he wasn't sure whether he should leave or not. I didn't mess with him anymore and went to work on the car fire. I put my air-pack on and went to help Rick and Tony.

When the fire was out, Rick told me what happened and I laughed about it. I wasn't about to chew him out about it. I just told him to be more careful next time. I then got on the radio and told fire communications to send us a tow truck and the police. I also informed them of our new location. I thought about how much fun that would be to explain in the fire report.

When we got the call, we were assigned a fire channel on the radio. Bullet forgot to change his radio settings before he talked on it for the first time. We had been out there for twenty minutes and the tow truck hadn't showed up yet. I told Bullet to call fire communications and get an estimated time of arrival for the tow truck. They came back on the radio and said, "Pumper 18, you're supposed to be on fire channel C Charles. What channel are you on?"

Bullet said, "You heard me, you find me." If fire communications talked to him on the radio, they knew exactly what channel he was on. There was no need to try to embarrass him like that. He just turned the tables on them. Bullet doesn't like to be embarrassed by anyone, especially by someone he doesn't even know. It wasn't right for fire communications to do that. All they had to do was ask him to change to the fire ground channel but the guy on the radio had to try to be funny. It's okay, though. The guys an asshole and everyone knows it.

We cleaned everything up when we got back to the station and I did the fire report. It took me awhile to find the right wording to use in the narrative of the report. By the time the fire was out, the cable on the parking brake had burned through. I just said the cable had burned through and the car rolled down the hill. The only damage the fire caused was to the car and the grass in the park. The car rolling down the hill caused no damage.

After the report was finished, I went to the kitchen for dinner but didn't even get my hands on a plate before the tones went off again. We were dispatched to the coliseum annex where they were having a home furnishings show. Vendors from all over the country come here twice a year to display their products. Everything from household furniture to dishes to accessories. When we arrived, everyone was standing outside waiting on us to arrive.

We walked in and couldn't see any fire. There was a lot of white smoke but no flames were visible from the front of the annex. The crews from Engines 18 and 15 started walking up and down the aisles trying to find the source of the smoke. After a few minutes, Ted Jones, a firefighter from Engine 15, came on the radio and said he had found the source. The fire was out but there were some lamps still smoldering. We found Ted and saw that there were over fifty lamps plugged into three outlets. The lampshades were all melted and there was a ton of smoke coming from them. The wires were melting and some of the lamps themselves were burned.

I took a couple of large garbage cans sitting in the aisle and started throwing the lamps into them. Rick and I took the first load out and the vendor who owned the lamps went nuts. He was pissed that we were dumping his lamps in the middle of the parking lot. I told him to calm down. I said, "Look, the lamps are history. We're just getting them out of there because we need to get behind them and check for damage."

He started stomping around and yelling, "I don't know why you have to put me out of business when everybody else gets to keep right on selling. I'm out of business."

I just turned and walked back in with Rick and we started loading the trashcans up again. I didn't see any reason to take the pails off the ladder truck when we had these cans here. Along the way, we each grabbed another trashcan and we were able to take the lamps out quicker. Once all the lamps were out in the parking lot, we were able to get to the floor and check for damage. The building itself wasn't damaged, but we counted twenty-five extension cords running from three electrical outlets. I couldn't believe that the show had gone on for three days with no problems. I went back outside and found the vendor. He asked, "Well, did you find the cause of the fire?"

I looked at him for a moment before I said, "Yeah, I found the cause. And you're it."

"What the hell are you talking about?"

"When you use three outlets to run fifty lamps, there's going to be problems."

The manager of the coliseum annex came up and asked, "Did you say fifty lamps?"

I said, "Yes, sir."

"Running off of three outlets?"

"Yes, sir."

He looked at the vendor and said, "I told you that you wouldn't be able to run all those lamps. That explains why you kept blowing the circuit breakers. You moron, you could have killed all of us!"

The vendor just stood there giving me dirty looks. I looked at him and said, "Don't look at me. I didn't cause this. You did."

Once we finished everything inside, we went to get on the truck and head back to the station. As we were walking across the parking lot, the vendor was talking to the police. As it turns out, he was being cited for all of the trash in the parking lot, the fire, and several other violations. He looked at me and I waved to him as we pulled out of the parking lot.

I decided to eat dinner and do the report later. I was hungry and I didn't care about the report. When I typed the report, I made sure that I accounted for every single light, extension cord and every bit of damage. I knew the guy would be pissed that we dumped all of his lamps in the parking lot and I decided to cover myself in case he decided to sue. I talked to the guys from Engine 15 and they said they would note all of the damage in their report and the fact that none of the lamps were salvageable. Once I finished the report, I decided to go to the classroom.

When I went in, Bullet was stretched out in a recliner sound asleep. He has been working two part-time jobs to pay for his honeymoon when he marries Jennifer. He hasn't been getting much sleep lately and any time he has a chance to sit down for more than five minutes, he's out like a light. When he goes to sleep, he's tough to wake up. It seems that the more noise you make, the better he sleeps. But make a small noise and he wakes up like there has been a shot fired next to his head.

Bullet has been known to sleep through some of the damnedest things. He has been carried out into the parking lot while asleep in his bed. Once, when we tried to carry him out, we dropped him. He didn't wake up either time. He has slept through several fights in the dorms of fire stations. We have made loud noises to wake him up and he never does. One reason I think he sleeps so soundly is because he snores. He could snore the siding off a house. I have never heard anyone who can snore that loudly and still sleep through it. He has even slept through gunshots.

Station 23 is located in one of the worst sections of town. Bullet was assigned there ten years ago when there were several drive by shootings in the neighborhood. One night, the shooters decided to fire into the station. They shot through the windows, then sped away into the night. All of the firefighters at the station rolled out of their beds and onto the floor except for Bullet. His bed is right beside the window and he never moved. He was lying there with glass all over him and he never woke up. Once the other firefighters were sure the shooters were gone, they went to check on him. He was still sound asleep. They weren't sure if he was dead or wounded. When they started checking him over, he woke up and started fighting. He had no idea what they were doing hovering over him like that. Once they explained what had happened and Bullet saw the glass all over him, he calmed down. I have never seen anyone who could sleep as good as him.

After watching a little TV, I called Teri to tell her goodnight. She could hear Bullet snoring in the background and started laughing. I gave Teri my love and decided to head to bed. I thought about waking Bullet up so he could go to bed, but I decided to let him sleep. I figured if he had slept this long in that chair, he was fine for the night. I hoped that if we got a call that night, he wouldn't get lost trying to find the truck. When the tones go off and you're in a different location than your bed you have a tendency to get confused. I decided that if the tones did go off, I would start yelling for him.

We were able to sleep all night with no calls. It seems that the night calls have slacked off here lately. I'm not complaining, just worrying. It seems that every time the midnight runs slow down, there is one hell of a call that will get the ball rolling again. But, like I said, we had a quiet night. At 7:00, I changed and went to the kitchen for coffee. When 8:00 rolled around, we changed shifts and I went home. Since I didn't have to go to my part-time job, I decided to go home and take care of some things around the house.

I worked on Danny's cane and took it easy all day. I didn't see any reason to kill myself on my day off. I then put Mojo and Jake in the car and went to pick the kids up from school. They were surprised to see the dogs but we had a good ride home. We stopped by the park on the way to let the dogs run around for a while. The park has a couple of fenced in lots that you can take your dogs to and let them run loose. There is an eight-foot fence around each lot and they are pretty big. The dogs love to be able to run around and Joey and Nikki love to chase them. I did remember to bring along a couple of tennis balls so the dogs would have something to play with.

After the dogs were tired out, we went home and found that Teri was able to get off work early. She decided to call it a day and we went out to dinner. The kids wanted pizza so we went to a local pizza restaurant and we had a big dinner. We went home and the kids did their homework. Once they were in bed, Teri and I decided to turn in early and spend some quality time together. I like quality time. Very much.

The next day, Teri let me sleep in while she got the kids ready and took them to school. She took the day off and we spent most of it laying around the house. There wasn't much to do because the house and yard work was done earlier in the week. It's nice to be able to sit around doing nothing. I just wish we had more time to spend together. We stayed out on the deck that afternoon since the weather was cooler. Fall's coming. Usually this time of year is hot, but we had a good rain last night, and it cooled things off for a while.

We picked the kids up from school and headed back to the house. We stayed out on the deck and the kids did their homework. It was so nice out there that we decided to have dinner out there as well. I cooked burgers on the grill and we made an evening of it. We had the music going and the Tiki torches were burning. It was nice out there. In fact, we stayed out there until it was time for me to go to bed. I had to work the next day, but they didn't. All in all, I think it was a good day. Of course, it would have been better if I didn't have to go to work on Saturday, but it's a living.

Chapter Nineteen

CPR in progress. When you start the day counting taters, or performing CPR, it's not going to be a good day. I don't know why it is, but damned near any time the call comes in as CPR in progress, there is never anyone performing CPR.

When we arrived on the scene, a woman met us in the front yard and said, "I think he's dead."

I asked, "Where is he, ma'am?"

"In the living room, straight through the door."

When I reached the man, he was sitting in a chair and was lifeless. He was not breathing and had no pulse. I had Bullet help me get him to the floor and we initiated CPR. I set up the semi-automatic defibrillator and opened the man's shirt. He was 83 years old and had a medic alert necklace on that said he had a pacemaker and diabetes. I was able to defibrillate him once, but the defibrillator did not advise us to shock him again. We checked his pulse and continued CPR.

EMS arrived on the scene and they took over care of the patient and we assisted them any way we could. Eddie Shoaf, the squad captain, works part-time as a paramedic with the county and the EMS captain told him he could start an IV in the mans arm. The second man on the ambulance had not yet been cleared to start IVs yet.

We worked on the man for about thirty minutes in the living room and he never responded to treatment the whole time. We put him on a backboard so that CPR could be continued and loaded him in the ambulance. All in all it was a smooth operation and we worked as hard as we could to bring him back. We just could not save him. I really hated it for the woman. It turned out that she was not his wife; she was a fiend of his. He lived at the coast and was in town visiting her. He went out for his morning walk and when he came back, she said that he had trouble breathing. After a minute of sitting there, the man went into cardiac arrest and died in her living room.

Bullet and I went back into the house and cleaned up all of the trash from EMS. We made sure the house was secure. The woman decided to go to the hospital with a neighbor and I told her that we would make sure the house was locked up tight. We locked the house up and headed back to the station.

We were all quiet on the way back to the station. We don't like to lose a patient. I know that people die and that, in this line of work, people will die while you are working on them. There is nothing that you can do to stop it. But that doesn't mean that you have to like it. And that doesn't mean that you don't try to save them.

The hardest part of it is dealing with the family and friends of the patient. After you do all that you can do for the patient, you have to be able to shift gears and deal with the victim's family. I have never been good at that. That's why I told the woman that it might be a good idea if she went to the hospital. I hated to do it that way, and I hate thinking like this, but I knew that if she left, I wouldn't have to deal with her loss. I know it sounds selfish, but I am never good at dealing with death and I never know what to say to someone who had lost a loved one. If they leave and go to the hospital with the patient, I know that I will not say the wrong thing.

I did the reports when we got back to the station. It's about half an hour of paperwork and I don't like doing it, but it's part of the job. It's not that the paperwork is hard to do, it's just that it is so redundant. I wish the EMS people and fire departments would get together and make one report that would cover all calls. I know that it will never happen, but I can dream.

Once I finished the reports, we went out in our first call territory and started checking hydrants. There's not much to it. We put a gauge on one outlet of the hydrant and check the pressure. Then we take off the other caps, oil all threads, and put the caps back on. There's not much to checking them in the fall. When we check them in the spring, we have to flow each hydrant as well as checking pressure and oiling the threads. A little more work but that's not too bad, either. It gets us out of the station and we don't have many classes when it's time to check hydrants. It's a good thing, too. It takes us the longest to check our hydrants. We have the largest territory and, even though we split the hydrants evenly between three shifts, it still takes about four weeks to check them.

We checked hydrants that morning for two and a half hours. We went back to the station to get some lunch and sit in the air conditioning for a while. As we were about to head out to check more hydrants that afternoon, we were dispatched to an injury from a fall, possible broken leg.

When we arrived on the scene, a woman was in hysterics in the front yard. She was yelling and screaming that her daughter was hurt and lying on the ground around back. We rushed around the house and found the girl, who was four years old, laying underneath a broken swing. As we started checking the girl out I asked the mother, "How did she get hurt?"

Through her tears, the mother said, "We were swinging together. I had her on my lap and for some reason the swing broke and we flipped over. I landed on her leg."

I looked at the mother as she spoke. She was a pretty big woman, weighing over 250 pounds. When she said "for some reason the swing broke," the only thought that I had was I knew why the swing broke. A child's swing is not made to hold an ass that big. She kind of looked at me funny when she noticed I was staring at her. I guess if she could read minds, I would have been in big trouble.

The girl turned out to have a broken femur. It was broken halfway between her knee and hip and we were worried that the bone could have severed the femoral artery. If that happened, she could bleed to death right in front of us. All we could do was hold traction on her leg and wait for EMS to show up. When they arrived, we put her leg in a hair traction splint. It has a strap that goes around the ankle and when it is put on, you can turn a knob where the strap is attached and pull the bone back into alignment. It sounds like it would be painful to wear, but once the bone is in alignment and you stop turning the knob, the pain eases up and the patient is more comfortable.

We baded her in the truck and let the mother ride in with her. When you have a conscious child in the ambulance, the paramedics will usually let a parent ride in the back with them to help keep the child calm so the paramedics can work on them. Ordinarily the paramedics don't want anyone else in the back, but it's hard to work on a screaming child who is in pain and is scared of you. When the parent is with the child, they have a sense of security because they can see mommy or daddy and they can assure the child that everything will be okay.

I decided to do hydrants for a few of hours before going back to the station and doing the report. I wanted to get the hydrants out of the way as soon as possible. The quicker we do the hydrants, the quicker we'll be done. I don't mind checking hydrants, but it's a good feeling when we get them done.

We got back to the station that afternoon and I did the report on the call for the little girl. Once I was finished with the report, I had to enter the hydrants we checked into the computer. I'm getting pretty good at entering them. You just have to find them on the computer and type in the pressure reading. Once I was done, it was time or dinner.

After dinner, we were sitting around the kitchen shooting the breeze when Jack Walters from EMS came by for a cup of coffee. We noticed that his new partner, James, wasn't with him. The woman riding with him, Melissa Franklin, has been with the county for several years and is a damned good paramedic. Jack told us about James.

Earlier in the day, they were at the hospital and Jack was filling out paperwork on a couple of calls they had that morning. While they were there, a private ambulance company came to pick up a body to take it to the medical examiners office. EMS and this particular ambulance company have had a running feud for the last two or three years and James wanted to have a little fun. When the EMTs from the private ambulance company loaded the body in the back of the truck, they decided to go get some coffee before they left. Jack and James took the body out of the back of their ambulance and James got in the back, got into a body bag and zipped it up. Jack hooked the bottom two straps and just laid the top strap across James' chest.

The EMTs came back to the truck and took off for the examiner's office. On the way, James started moaning from the back of the truck. One of the EMTs said, "What the hell was that?"

James sat up on the stretcher and started flailing around like he was trying to get out of the body bag. The EMT in the passenger seat grabbed an old metal bedpan and whacked James in the back of the head. James went out cold and the driver stopped the truck. They checked the bag and found James, unconscious. They turned the truck around and headed back to the hospital. When they arrived, they opened the doors and pulled the stretcher out. They tipped it and dumped James on the sidewalk beside the ER doors. They found the body bag they were supposed to be transporting, checked the body inside and put him back on their stretcher and left James lying there. Jack walked out and asked the driver where his partner was. They pointed to the half open body bag lying on the sidewalk and the driver said, "Next time he tries something like that, tell him we'll beat the shit out of him with that bed pan."

Jack walked over and James was beginning to regain consciousness. Jack looked at him and asked, "Are you okay?"

James sat up, rubbing his head, "I guess so. I've got a killer headache."

"What happened when they found you?"

"I don't know. All I know is I was sitting up on the stretcher yelling, 'Get me out of here!' Last thing I remember was a loud clang and then everything went black."

We were rolling with laughter. I asked Jack, "Did James go home with a headache?"

He said, "No. He went home with a concussion. And one hell of a bump on his head."

That just made us laugh even harder. The best part was when Jack went out to his truck and came back with another metal bedpan. We took it to the engine room and started working on it. We painted it with different colored paint in a camouflage pattern. We added a strap to the opening and glued some twigs to it. On the front we painted the words "EMT Hard Hat" and on the back we painted a bulls eye. We then put the bedpan/helmet in a box and wrapped it up with a big bow. We also put a little poem in with his "present" that said,

> Roses are red, violets are blue
> This is a helmet to protect you
> When you are riding in the bed of a dead man
> It will protect you from flying bedpans

Jack took the present and headed over to James' house. After an hour he came back and told us that James didn't really appreciate the humor in the gift. He threw it at Jack and slammed the door in his face. Jack said he took it back to the base and put it in James' locker so he would see it first thing when he came back to work on Tuesday morning. What a way to start your day.

We lucked out again. Another quiet night. After calling Teri and telling her goodnight, I went to bed and I was able to sleep all night. It's nice when that happens but eventually something will come up and bite you in the ass. Once that happens, it seems that you go several weeks without a decent night's sleep while one duty. I guess I'll enjoy this while I can.

After getting off work, I went home and found everybody in bed. I figured that Teri would want to go to church but she was still asleep when I got home. Of course, that didn't last too long because the dogs were so excited to see me that they ended up jumping all over her on the bed. Not a very nice way to wake up. After giving her a kiss, I asked, "Aren't we going to church today?"

She said, "Joey doesn't feel good so we're just going to stay home."

"Okay. You go back to sleep and I'll get the dogs out of here. I'll be outside or in the shop."

"Okay."

I kissed her and went to the shop. I worked on Danny's carving. I've been working on it for a while and I'm close to getting it finished. This will probably be a one of a kind cane because if I ever get this thing finished, I'll probably never do another one like it. But it is turning out better than I thought it would. And I am nearly finished.

I worked out there for a couple of hours before anyone came to see me. Nikki came out to see what I was doing. She told me that Teri was cooking breakfast and it would be ready in a few minutes. I cleaned my mess off the floor and put the carving back on it's shelf. You know that you have been working on one project too long when you give it it's own shelf to sit on.

I went into the house and Joey was laying on the couch. He wasn't feeling good at all. He had a temperature and he was achy all over. I guess it's a good thing that Teri is a nurse and can take care of him. I went to the kitchen and Teri had been cooking up a storm. We had one hell of a breakfast. It's just too bad that Joey couldn't enjoy it. But Teri had been promising to cook a big breakfast for me for a long time now and she decided that today would be the day.

Joey decided to go back to bed. Nikki and I went to the shop. Nikki wanted to make some shelves for her room like Joey and I made a few weeks before. I cut the wood and we glued everything together. While she waited for the glue to dry, she helped me clean out some old scrap wood that I couldn't use for anything. Usually when I get a bunch of scraps together, I take them to a friend's house and I give them to his son. He loves to play with the blocks of wood and he now has quite a collection.

Once the glue dried, I let Nikki paint the shelves however she wanted to. I went back to the cane once again and did a little more carving on it. I actually finished the carving and all that was left was a little sanding and actually putting the cane together. It won't be long until I can finish this thing up and move on to another project.

After spending most of the day in the shop, Nikki and I cleaned up and went back to the house. The minute I walked in, the phone rang and it was my part-time boss, Rick Jackson, and he informed me that they would be closed again for another week. I figured that I would stay home and finish the cane tomorrow. I didn't have anything else to do and I would be able to get everything glued up and screwed together. What the hell, I'll take another day off.

Rick told me that it didn't look too good for the company. The owner was seriously thinking about just closing the business and retiring. It didn't bother me one way or another. I didn't have to work part-time. I basically was working there to have something to do. When I started working there, I wasn't into woodworking like I am now. I make furniture and I carve. I now have something to fill the days I'm off from the fire department and I don't need the money that bad. I figure I can start carving and sell the what I make. Then again, I may just become semi-retired.

Teri and I turned in early after we tucked the kids in. Joey was feeling a little better and I would be able to stay home with him tomorrow if he didn't go to school. I didn't think he would, but you never know. I don't know if he ate something that made him sick or he picked something up at school. Either way, he wasn't feeling good and all he wanted to do was sleep.

The next day, Teri went to work and Nikki went to school. Joey and I hung around the house and took it easy. He was feeling better but was he was weak. He was able to keep what food he ate down today and I pushed the fluids in him. The only problem he had today was the fact that he had to get up every ten or fifteen minutes to go to the bathroom. Teri called a few times to check on Joey and make sure he was resting. The last time he stayed home from school, he was sick but he also spent most of the day on the computer. Today, I didn't let him get on the computer. I made him rest and take it easy.

Nikki went to the YMCA and Teri picked her up on the way home from work. On the way home, they stopped and picked up some burgers. We had dinner and Nikki went to do her homework. She had checked with Joey's teachers and they didn't have any homework for him. Joey turned in early and Nikki went to bed shortly after that.

Teri and I stayed up and watched a movie. We turned out all of the lights and settled in on the couch. It was a pretty good movie; afterwards, we went to bed and continued our cuddling session. But it was getting late and we both had to work the next day. We fell asleep spooning each other. It's a nice way to fall asleep but when we do that, Jake gets pissed because we're in his spot on the bed. Sometime in the night, he managed to force his way between us and take his spot back. He is persistent.

Chapter Twenty

We went out to check more hydrants today. We did forty-five hydrants this morning and forty-eight this afternoon. It was a nice day outside so I didn't mind it too much. It gets a little boring doing the same thing over and over all day but at least we were outside. Beats sitting around the station with nothing to do.

After we finished the afternoon session, we headed back to quarters and I entered everything in the computer. It didn't take long. Once I was finished, I went to the kitchen to see if dinner was ready. I had worked up an appetite and Bullet was cooking. When I walked in, they were just getting ready to call everybody on the intercom system.

Bullet out did himself. He can make something great to eat in such a short time. We had a good meal and then we kicked back to watch a little television. I realized this was the first chance I had to relax all day. It actually lasted the entire evening. We watched a couple of movies on cable and then I decided to call Teri and tell her goodnight.

After I gave Teri my love, I headed to bed. I went to sleep pretty quickly but it didn't last long. At 2:30, we were dispatched to a reported house fire a couple of miles from the station. When we made the turn from the front ramp to the street in front of the station, I could see the glow from the fire. I knew this was going to be a good one. I called fire communications and told them that I could see flames from the station. When you are that far away and can see the glow, it usually means the fire is rolling.

We arrived on the scene to find that the fire had vented through the roof of the house and was rolling out the upstairs window on the west end. I had Rick catch the hydrant and Tony and I got our air-packs on and took a line to the door. By the time we were ready to enter the house, Engine 15 arrived and backed us up on another line. We headed up the stairs to hit the fire when Chief Williams arrived and took command. Bullet told him what had happened so far and that we were inside getting ready to hit the fire.

Once Rick and I reached the top of the stairs, I called for water. Bullet opened the line and we hit the ceiling to knock the fire back. We were able to knock it back enough to get to the fire room. Once inside, the hose got jammed between the front door and the newel post at the bottom of the steps. We couldn't advance the hose any further until it was freed from the bottom of the steps. Tony had the nozzle and was working like hell to keep the fire off of us as I got on the radio and told command that we needed somebody to help us with the hose.

It took about four calls before Chief Williams finally realized what I was talking about and sent another three men in to help with the hose. Once we were able to advance it, the fire had built up pretty high and another few seconds and we would have been toast. Rick said he needed me to take the nozzle. He had worked himself to near exhaustion and needed a break. I took the nozzle and started working it on the ceiling. I was able to find the seat of the fire in the room. I hit it and the smoke and steam vented through the hole in the ceiling. I knocked the fire out in the room pretty quick but there was still a large amount of smoke.

Rick and I worked the nozzle back toward the closet and found a lot of fire in there, too. It was a large, walk-in closet and everything in there was history. We were hitting the fire in the closet when the crew from Engine 14 came in and started hitting it in the attic. Oddly enough the fire in the attic was contained to the west end of the house. There was a considerable amount of smoke but the only fire was over the room Rick and I were working in. It hadn't spread any further.

It took awhile but we were able to get the fire knocked out. There was a ton of stuff in that bedroom. Apparently, the owner of the house was a magician because there was a lot of magic props in the closet. There was a large metal desk in the bedroom and every time we pulled a drawer out, the contents were burning. The window on the end of the house was gone so when we came to the contents of the room that weren't salvageable, we tossed them out the window. We did a secondary search of the house and found no victims, then we salvaged what we could.

Once the fire investigators came in, we moved outside to take a break. I may not agree with Chief Williams on a lot of things, but one thing I do agree with him about is rehab. When we walked outside, the rehab truck was already there with coffee and drinks. I went over to Chief Williams and he told me what the owner of the house said.

He said, "The fire started in the bedroom you were in. Apparently she was drinking and smoking and dropped her cigarette on the couch. She tried to put the fire out herself but she couldn't do it. She took a five-gallon bucket from the kitchen and filled it with water. She ran up the stairs and dumped it on the couch. She said it didn't put the fire out completely so she ran back downstairs to get another bucket of water."

I asked, "Why the hell didn't she just go to the bathroom next to the room?"

"I don't know. Maybe because she's a drunk moron."

"We found a bunch of toys up there. Does she have kids living here?"

"Yeah. They were asleep in the next room. Once the room was too hot for her to go into, I guess she decided to finally get her kids out of there."

The stupid things people do when they get drunk. She could have got her kids out of there and called the fire department and the fire would have done a lot less damage to the house. But instead, she tried to put the fire out herself. She was stupid as hell not to get the kids out of the house first. She put them in danger because of her drinking. Not only that, but when you have a bathroom with a tub in the very next room, why the hell would you keep running downstairs to the kitchen to get a bucket of water. I felt sorry for the children because they lost a lot of their toys. But I didn't feel sorry for the mother at all. It was her drinking that caused the fire and it was her stupidity that allowed it to nearly destroy her house and kill her children.

While we were standing out there, one of the police officers, Mike Richards, came over to get a cup of coffee. He was doing traffic control for us and we were more than happy to give him all of the coffee he wanted. When he walked up, I noticed that he had a black eye. I asked, "What happened to the eye, Mike?"

He said, "Got in a fight with a perp a couple of hours ago."

Laughing, I asked, "Who won?"

"Hey, you don't know what I was up against. I fought that guy for ten minutes before I could get the cuffs on him."

"Why didn't you pepper spray him?"

"I sprayed him three times! Every time I sprayed his ass, he just ate it up and came back for more. Come to find out, when I was filling out the report, I asked him what he did for a living. He told me he was a chemical tester with for Marines."

We laughed. But I do have to admit that Mike allows us to meet some of the best fighters in town. He isn't quick to spray someone, but when he does, he usually brings them to our station to get their eyes washed out.

That's one thing that Bullet loves to handle. We take a garden hose, take the nozzle off, and flow water over their face to cool the burning. At least, that's how we are supposed to do it. Bullet usually changes the angle of the hose so that it's shooting water straight up their nose. He has a tendency to play with them a little. He'll move the hose away from their face so that they have to lean over more and more. He has even made a few of them fall over from trying to get their face in the stream of water. The water only cools their face for a few seconds, then the burning comes back. It's only a temporary cure.

We finished working the fire ground and headed back to the house. We cleaned up and loaded the truck. I went and took a quick shower so that I could stop offending myself. As I was getting ready to lay down, the tones went off again and we headed for the truck.

We were dispatched to a traffic accident with possible pin-in. When we arrived on the scene, we found a pickup truck sitting on the edge of a creek. The rear of the truck was in the water. The front was just off the edge of the street. The driver hit a tree and the rear of the truck whipped around and mired up in the mud in the creek.

We could see two victims in the truck and it didn't look good. I went to the passenger side of the truck to check on the guy hanging out the window and Rick took the driver's side. I checked for a pulse and he was dead. There was no way that we would be able to save this guy. I looked at him and the frame of the window of the door had cut half way through his body from the back, tearing everything in its path. I looked over the hood of the truck and Rick told me that the driver was dead, too. There was no way to get him out of the truck because the engine had pinned his legs in. We then switched sides of the truck to check each man again. I checked the driver and Rick checked the passenger. We did this just to be sure that the victims were, in fact, dead.

We stood around waiting for the tow truck to arrive which, of course, seemed like it took forever. Once it arrived, we hooked to the frame of the pickup truck and hauled it out of the creek. When we had it on level ground, we started the process of extricating both victims. We worked on the passenger first.

The way he was positioned in the window of the truck, we weren't sure if he nearly flew out of the truck or if he was riding on the windowsill. Once we got the door open and the window post out of his back, we then had to cut the door in half. The door had wrapped around his right leg and we had to cut the door away. It took us awhile to extricate him. By the time he was freed, the medical examiner arrived with his truck. We put the passenger on the coroner's gurney and rolled him to the truck.

While he took the victim to his office, we extricated the driver. It didn't take as long for us to get him out of the truck. Once we forced the door open, we used a hydraulic ram and moved the motor off of his legs. When the motor was moved, we were able to pull him from the truck. Both legs were broken in several places. They moved in ways that looked painful until you realized he was dead and couldn't feel it. When we pulled him all the way out, I saw that his eye was rolling around on his face. He hit the ceiling of the truck with so much force that it not only broke his neck, it forced his eye out of its socket. That was a first for me. I had never seen anyone who so little facial damage with an eyeball rolling on his forehead.

After the coroner came back and we loaded the driver in his truck, the parents arrived. They lived in the neighborhood and were worried about their son when he didn't come home on time. When they saw all of the lights on the police cars and fire trucks, they decided to go see what happened. When they saw their son's truck so badly damaged, the mother went nuts. She started screaming and struggling to get into the coroner's truck. The police had to drag her away so the coroner could leave.

Fortunately, the police dealt with the family so we didn't have to. After the kids were gone, we waited for the tow truck driver to hook the pickup truck up. While we were waiting, I looked inside at the damage. Then I noticed the speedometer was stuck on 95 miles per hour. The wreck happened in a big curve and that was way too fast for the curve. I guess the driver just couldn't handle the road. When the tow truck left, we washed the street down and put our equipment up. We headed back to quarters while watching the sunrise. I was thankful that the past few nights were quiet, but like I said before, when it's that quiet for that amount of time, the nights get going again with a bang. Tonight, we were hit with a double whammy.

I decided not to try to go to bed when we were back in the house. There was no point in going to bed when you had to get right back up in an hour. We just sat in the kitchen drinking coffee and waiting for shift change. Finally, the other shift took over and we were able to get out of there. I went home and had a melt down in bed. No one was home and the dogs never mind it when I take a nap. They were more than happy to join me.

I got up around 2:00 and I decided to get something to eat. The only problem was that there was nothing in the house I wanted. I decided to go to the mall and get something in the food court. I needed to get some new running shoes anyway so I figured I would get something to eat and do a little people watching at the same time. It never fails. There's always something going on at the mall to watch.

Mostly it's the young guys there that put on the best show. I have never been able to understand why anyone would put a pair of pants on that is three sizes too big and barely hanging on their butts. That and the baseball hat either turned sideways or backwards. Sometimes it seems that they are having a contest to see who can look the most ridiculous. If they aren't wearing a hat, their hair usually looks like the opened the hood of their car and rubbed their head around until there is enough grease to do a lube job.

We had a call one afternoon to the mall that turned out to be an injury from a fall. One of these guys had been walking around the mall like Joe Bad-ass. He had the whole look going, too. He was wearing three shirts and a baseball hat turned backwards. He had sunglasses on and there were about twelve piercings in his face. And, of course, his pants were way too big. He fell when he was getting on the escalator and his pants went down too far and tripped him.

He kept asking, "Am I going to be all right?"

I said, "If you would wear clothes that fit you, you wouldn't have to ask that question."

He said, "It's the style."

"Well I guess now you're a fashion wreck. Style is one thing, but when you look this ridiculous, it's not good."

He wouldn't listen but I didn't care. I wasn't the one going to the hospital with a broken leg. Besides, I was able to go down the escalator without falling. I never have understood why anyone would want to look like a child wearing their parent's clothes. It doesn't make sense to me, but then again I've always been a T-shirt and jeans kind of guy.

I got my food and sat down in the food court to watch the show. There wasn't a whole lot going on at first but the show got better when more and more kids started arriving after getting out of school. The kids put on the best show. The guys are there trying to act tough for the girls and the girls are there putting on too much makeup and trying to ignore the guys. It gets even more interesting when there is more than one guy going after a girl.

Today's show featured two guys that were trying to smoke in the food court. Smoking isn't allowed in the mall and they wanted to be rebellious. They sat down a couple of tables away from me and lit up. Two old women sitting at the next table told them that they weren't allowed to smoke in the mall. One of the guys looked at her and said, "Mind your own fucking business, granny. I'll do whatever the fuck I want, whenever I want."

The old ladies just sat there with shocked looks on their faces. Then a woman sitting at another table told him that he really needed to put the cigarette out. He told her, "Shut the fuck up, bitch."

At this point, I had had enough. I walked over and said, "Look, pal. Put the cigarette out and apologize to the ladies."

He looked at me with a sneer on his face and asked, "And if I don't?"

I reached down and grabbed his nose ring and said, "If you don't, I'm going to turn this nose ring into a nipple ring. Now start talking."

He turned to the old ladies at the first table and said, "I'm sorry."

"Tell them you're what you're sorry for."

"I'm sorry for talking to you that way."

I then pulled his head around so that he was facing the woman at the other table and said, "Now for the next apology."

"I'm sorry for talking to you in a bad way, too."

"Now, put the cigarettes out." They dropped them in their drinks. "Now, get out of here."

After dealing with the shit-head twins, I headed over to the sporting goods store to get my shoes. I found some good ones at a decent price and bought them. I had a little time left before going to get the kids at school so I did a little more shopping. There is a great woodworking supply store in the mall and I went in to check out their carving chisels. I bought a few chisels and a couple of books.

By the time I finished buying the tools, it was time to go get the kids. When I got to the school, the kids were outside. There was a fire alarm and the teachers had the students lined up in the yard in front of the school. As I pulled up, Lionel Stokes, the A-Shift captain on Engine 3 was coming out of the school and heading back to the truck. I called him over and asked what he had.

He said, "Nothing to it. Just a false alarm. Hey, I heard you guys had a rough night last night."

I said, "You heard right. Not only a house fire but a double fatality wreck."

"Where was the wreck?"

"Creekside Drive. Pickup truck versus tree. The tree won."

"Damn, that's rough. How old?"

"I'm not sure. They didn't look that old and one had a college ID in his wallet. It was pretty grim. And then the parents showed up. But fortunately we already loaded both kids in the coroners wagon."

"Well, we gotta go. You take it easy."

"Take it easy. Stay in the house."

"I hope we do."

The kids went back into the school and picked up their books. They came back out and we headed home. The kids started their homework and I went to my studio. I started working on a couple of pieces of music. I may not work every day but I do have something to do when I'm not at the fire station. I figure that it's better to have something to do than to veg out in front of the television. Besides, I like making music. Good or bad, I do enjoy it.

Time got away from me while I was in the studio. Teri walked in and kissed me hello. I didn't realize it was time for her to be home. She didn't mind it too much, though. She went to the kitchen and made dinner while I sat in the studio and played for my supper. She likes to hear me play. Even if I am struggling to learn a piece, she still enjoys it.

We had dinner and I told her about the night before. Usually all she has to do is ask how my night went. If it wasn't bad, I just say, "We had a good night." If it was bad, I usually say, "It went." She understands. I don't talk about the bad calls in front of the kids very much. It's one thing to talk about a house fire, but I wouldn't think of talking about the wreck in front of them. I did talk to her about it later, though. I just needed someone to talk to and get it out of my system. At the station, we have each other to lean on. At home, I have Teri. She is the anchor of my life and she helps me in so many ways. All I have to do is talk to her and it always makes me feel better.

We put it to bed early that night. While we were laying there, Teri gave me my list of womanly chores for the next day. I don't mind helping out with the housework but I do wish she would write these things down. I have a tendency to get distracted easily and if I have a list I do better. I got my list of chores and then fell asleep holding her. It didn't take Jake long before he was moving me over so that he could take his spot on the bed. I just rolled over and went back to sleep.

The next day, I miraculously remembered everything that Teri told me she needed me to do. I even did a couple of things that she didn't ask me to do because I couldn't remember if I was supposed to do them or not. When I miss a chore or two, she doesn't get mad. But I still feel like I didn't hold up my end of the bargain. I just want everything to run smoothly in the house so I like to help out where I can.

It didn't take long for me to finish up. I went out to the shop to put the cane together. I put everything together and sanded all of the joints. Then I put the first coat of finish on. It looked pretty good, even if I do say so myself. And it will be a one of a kind. If I never make a cane with this much detail again it will be too soon. But this one is for a good man, Danny Steele, who had his career shortened by a fall. I don't know if he will ever use it, but I wanted to do this for him because he helped me out when I first came on the fire department.

Going through training, I was having a hard time learning fire behavior. I just couldn't picture it in my mind. Danny took me into the house on a practice burn and explained how the fire would travel after being lit and how it would act while burning.

He said, "Fire is a living, breathing animal. It needs to eat to live. It needs to breathe to live. If there is no fuel and no oxygen, the fire dies. It will build up the wall burning everything in it's path. Then it will travel to the next room looking for more food and oxygen. When it eats everything in that room, it will look for more food. It needs air and when it uses all of the air in here, it will go looking for more air. Just watch what it does."

He then lit the fire and I watched as the fire did everything that he said it would. While it was burning, he pointed out when it started looking for more fuel and when it started looking for more air. Once he showed me how it acted, I understood fire behavior. It was one of the best training sessions I have ever been to and I can only thank Danny for taking the time to show me. That's why I want to do something for him now. Danny is a good man and was a great firefighter. Too bad the fire department wouldn't do anything for him after his injury.

I don't understand why they couldn't make him training chief or give him a job in fire investigation. With his knowledge, he would have made a great investigator. But that's the life of a city employee. Once you can't do the job that you were hired for, they basically say, "Fuck you. We're done with you. You have to retire."

After finishing in the shop, I went to get the kids. This time, things were a lot calmer at the school. Joey came out first and got in the Jeep. It took Nikki a little longer to get out there. I knew why. She likes to talk and I knew she was in there talking to her friends. I didn't mind waiting for a few minutes. When she came out, she was in a good mood. She got in the Jeep and we went to the house. She was in a good mood because she didn't have any homework to do today. That always made my day when I was going to school. Joey, on the other hand, had quite a bit of homework to do.

When we got home, he went to his room and started working and Nikki and I went to the shop. I carried her shelves to the house and we worked on her room for a while. She doesn't like to clean her room but by helping her rearrange her room I was able to trick her into cleaning it up. When I pointed out the fact that she did all the cleaning, she got mad. But, it's her job to keep her room clean and she knows that.

Joey finished his homework by the time Teri came home. I called out for pizza and we put a movie on. After eating and watching the movie, it was time for the kids to get their things ready for school. I went to get my uniform ready for work the next day and take a shower. I crawled into bed with Teri and the dogs and we cuddled for a little while. Then I fell asleep while listening to Mojo snore. Sometimes he can really saw logs. I had to move him a couple of times to get him to stop. Once I was asleep, I didn't hear him again until the next morning.

Chapter Twenty-One

We had a slow day today. We went out and checked hydrants for a while in the morning. Not much to it, but because it is so repetitive, it makes the time go by pretty slowly. We did manage to get forty hydrants out of the way, though.

While we were out there, we ran into Brian Walker, a captain on Engine 24, A-Shift. We came on together and we've been friends ever since. The thing is, Brian has a tendency to be a smart-ass. Case in point, last Wednesday. The Battalion Chief on A-Shift, Eddie Anderson, is known as a real hard case. He chews men out in front of the other men. He holds grudges for a lifetime. And if you get on his shit list, he'll make your life miserable for the rest of your career.

Last Sunday, Brian and his crew took the truck to the grocery store for food. On Sundays, most of the stations have a big breakfast but no one on Brian's crew had a chance to get their food on the way to work, so they went to the grocery store after shift change. They went to a store two blocks down the road from the station.

Apparently, Chief Anderson came by the station and couldn't find them. He went ballistic. When Brian left the station, he called fire communications and told them that Engine 24 would be in service in their first call territory. Obviously Chief Anderson didn't hear them go out because he was furious when Engine 24 returned to quarters.

Chief Anderson chewed Brian's ass out in front of the crew and an elderly couple that came by the station for directions. In short, Brian was told that every time he took that truck out of the station, he was to call Chief Anderson not matter what. Chief Anderson said, "I don't give a damn why you leave the station, you had better call me first!"

Brian said, "I couldn't believe he was chewing my ass in front of the crew and two civilians. But I got his ass back."

I asked, "What did you do?"

"Well, we didn't leave the station again all day Sunday. But we rode like hell on Wednesday. That morning, we had to go to the shop. I called him and said, 'Chief, we're going to the shop.' He said, 'Okay, that's fine with me.' Later on, we had a ride and I called him and said, 'Chief, we have a call and we're taking the truck out of the station.' I didn't give him a chance to say anything; I just hung up and rode the call.

"We had two more calls that afternoon and four calls that night. Every time we rode, I got on the phone and said, 'Chief, we have a call and we're leaving the station.' I never gave him a chance to respond. I just kept hanging up on him."

I was rolling. I said, "I know that had to piss him off."

"Oh, it did. The last one was at 3:30 yesterday morning."

"Wait a minute. You mean to tell me that you held up a call to call Anderson on the phone?"

"Hell no. I used my cell phone. The best one was when we rode a fire call that evening. I called him on the cell phone in the chief's car and told him we were leaving the station. He yelled, 'God damn it! I know you're leaving the station. You're right in front of me!'"

I asked him, "How long did he chew you out?"

"He didn't. He just called me yesterday morning and said, 'Forget I said a damned thing. Forget I told you to call me whenever you leave the station. Just quit calling me all of the time! Please!'"

"You know he's going to shit a brick when you call him again."

"No doubt. And the beauty part of it is, I have to call him tomorrow morning to tell him that we'll be covering a practice burn tomorrow instead of Engine 12."

I loved it. And I can think of no one on this department who deserves that kind of treatment more than Chief Anderson can. He feels that he deserves respect from the men on this department because of his position. The worst I was ever chewed out by Chief Anderson was the day I told him, "Chief, you don't get respect because of your position. You get respect because you earn it." He chewed my ass for nearly an hour. Of course he did it in front of the entire station and that just proved my point even more.

That's one of the things I respect most about Ray Jones, our former battalion chief. He isn't afraid to chew someone's ass out. He'll do it in a heartbeat. The difference is that he will not do it in front of anyone else. He gives us the courtesy of not embarrassing us in front of other people. Not to mention the fact that, once he is done chewing us out, it's over. Chief Anderson cannot do that. If he chews your ass out, he'll remember for the rest of his days. Chief Jones is different. Once it's over, it's over. And you will not hear about it again.

We finished the morning's hydrants and headed back to quarters. I had enough time before lunch to enter them in the computer. Then I went to eat lunch and kick back for a while. We sat around talking about what Brian did to Chief Anderson. That brought on a lot of horror stories of everyone's dealings with him. The only one who didn't have a story about him was Bullet. I turned to him and asked, "Bullet, haven't you ever had a run in with Anderson?"

He said, "I've come close to dealing with him once."

"What happened?"

"He was starting to chew my ass for something that wasn't even my fault. I cut him off quick and said, 'Chief, don't fuck with me right now. I'm not in a good mood and if you piss me off, I'll go to the truck, get my gun, and shoot your sorry ass right here and now.' I was just joking but I think he believed me."

Rick said, "Damn, I wish I would have thought of that."

I said, "Yeah, but Bullet's the only one psycho enough to do it."

Rick thought for a moment, then said, "That's true."

We did another thirty-five hydrants that afternoon. Rick and I took turns turning the hydrants and Bullet wrote down the pressure readings. We stayed out there for four and a half hours. Every once in awhile, some kids would come up to see what we were doing. We let them look at the truck and sit in the driver's seat. It turned out to be a fairly enjoyable afternoon.

We went back to the house and, after dinner and entering the hydrants into the computer, I settled in a recliner to watch a movie. We were watching a pretty intense horror movie. Mike Berrier loves to watch horror movies but he gets so wrapped up in them that he cannot hear or see anything going on around him. Our classroom has two rows of recliners sitting on tile floors. To keep the scuff marks down around the recliners, each one sits on a small, cloth rug.

In one of the more intense scenes of the movie, I nudged Bullet who was sitting next to me. I whispered, "Watch this." I then leaned forward to Mike's chair and grabbed his shoulder and yelled, "Boo!"

It scared the shit out of Mike. It scared him so bad that he jumped out of his chair. His feet were moving so fast that when he jumped up, his feet yanked the rug out from under his recliner without moving it. The rug went flying up in the air and Mike stumbled and fell in Chief Williams' lap. Chief Williams shoved Mike out of his lap and said, "God damn, Mike! If the movie scares you that bad, maybe we need to get your mama down here to hold your fucking hand!" We laughed so hard that we had to rewind the tape to watch the ten minutes that we missed.

After the movie, I called to tell Teri goodnight. I have never liked being away from her, especially at night. But it's part of the job and she understands. She doesn't like it any more than I do but she understands that I am doing a job that I love and she wouldn't ask me to change that. I gave her my love and went to bed.

At 1:30, the tones went off. We were dispatched to a neighborhood where we had just checked hydrants the previous afternoon. The call came in as an O.B. in labor. As I got on the truck, I went through the steps of childbirth in my mind. And then the thought hit me. I looked at Bullet and asked, "Do you think that we need to call Chief Anderson at home and let him know we're leaving the station?"

He laughed and said, "You do and he'll probably shoot your ass."

We arrived on the scene and the expectant father led us to the master bedroom, upstairs. I kept thinking to myself, "This guy looks familiar. I know I've seen him somewhere before." When we walked into the bedroom, the mother was laying on the bed in active labor. Her legs were spread apart and her breathing sounded like a freight train. Since she looked like she was having some difficulty breathing, I started to set up the oxygen bottle. Rick checked her for crowning. It's not the most dignified thing to do, but it's necessary. He raised up her nightgown and looked between her legs. He was looking to see if the woman was fully dilated and if the baby's head was starting to push through. Rick looked at me and said, "Cap, we've got a head!"

Bullet, who was standing in the doorway, turned and walked to the hallway. He started pacing back and forth and muttering to himself, "I don't do babies. I don't do babies."

I took the O.B. kit from the medical bag and started handing things to Rick. He looked at me with an astonished look on his face. He asked, "You mean I have to do it?"

I said, "Well, bubba, you're the one down there playing Johnny Bench."

The baby came out pretty quick. It turns out that this was the woman's second child. The whole time we were delivering the baby, her husband was running around like a chicken with his head cut off looking for the camera. He kept yelling, "Hold on, honey. Hold on until I find the camera." She didn't listen. In less than a minute, the baby was out.

As soon as the baby was delivered, the paramedics showed up. It figures. We do all of the work and they get all of the credit. A few seconds later, the woman shot a blast of air from between her legs directly into Rick's face. He looked at one of the paramedics and said, "I need you to take this baby!"

He said, "What?"

"I need you to take this baby. I gotta get some air. Take the baby now!"

The paramedic took the baby and Rick ran out of the room. Bullet looked at his face and saw that he was about to toss his cookies. Bullet pointed to a room off the hallway and said, "Bathroom's over there."

I just laughed to myself. Rick always did his best to be professional and not lose his cool. But all it took was one quiff from a woman and he damn near lost his lunch.

The father came back in with the camera. As soon as he saw the baby he started whining, "Oh, honey, you didn't wait for the camera."

I looked at him and said, "Well, we could put the baby back inside so you can get a picture."

She gave me a dirty look and said, "The hell you will!"

We cut the umbilical cord and packaged the mother and baby for transport. She hadn't delivered the placenta yet, but the hospital is such a short distance away that we didn't need to wait. As we loaded her in the truck, she asked one of the paramedics, "Do you think this is any of their first time?"

He said, "I don't know if any of them have ever delivered a baby before or not."

"No, that's not what I mean. Do you think it's the first time any of them has ever seen a woman's privates or not? Because, if it is, they are welcome to come see it again."

The paramedic didn't know what to say. He said, "Well . . . uh . . . Ma'am . . . I'm not sure . . . Probably not . . . Ma'am we need to get you to the hospital!"

I laughed my ass off as I closed the doors. I told Rick and Bullet what she said. Rick said, "Well, that's the first time I've ever seen one do that before! That's for sure!" He then spun around to Bullet and said, "And as for you, you sorry motherfucker! You were no help at all!"

Bullet said, "I've fished enough babies out of toilets and car seats that I don't need to do it anymore. Besides, you guys had it handled anyway."

The father walked over to thank us and that's when I realized where I had seen him before. That afternoon, he and his son had come by the truck to see what we were doing when we were checking hydrants. I said, "Look. The next time we're here checking hydrants and your wife is near her due date, tell her to deliver the baby while we're in the neighborhood. It's easier then than the middle of the night." He laughed and thanked us again.

We started to head back to the house but, before we could get there, we were dispatched to a hostage scene at a convenience store. We were to standby in case anyone needed medical attention.

Fortunately, we only had to standby for twenty minutes. Sometimes these things can take hours to work out. But the guy gave up and surrendered to the police. We were on our way back to the station when we were dispatched back to the scene to assist the police department's tactical team. We weren't sure what they needed.

When we got back to the scene, we found out that two members of the tac team had climbed on top of a building across the street from the convenience store. Now they had no way to get down. I guess it was easier going up than down. Bullet looked up there and started laughing. One of the police officers said, "You have a lot of guts to laugh at two men with guns, who can shoot you."

I said, "But do you have the stupidity to shoot the three men with the ladder that can get you down?"

"Good point."

We got them off the roof and finally returned to quarters at 3:45. Of course, with all of the paperwork, it was 4:30 before I got back to bed. In another two and a half hours, I would have to get up and get ready to go home.

After getting a little bit of sleep, I decided to take a shower. After getting in and soaping up, the tones went off. It never fails. Whenever you try to shit, shower or shave, the tones go off. Just par for the course, I guess. I rinsed and dried off as quickly as I could. Then I dressed and headed for the truck.

We were dispatched to a rescue call in our own neighborhood. When we got there, we saw a police officer standing on the sidewalk. As I got off the truck, I asked what we had. He pointed to the top of a tree and said, "Up there."

I couldn't believe it. There was a naked man sitting in the top of the tree. Then the officer explained, "His girlfriend told me that he was dropping acid all night, then climbed up the tree and stripped naked. I guess the LSD wore off because now he's afraid to come down."

I called the ladder truck out to assist us. Upon their arrival on the scene, they raised the aerial and Mike Berrier climbed up there to help him down. On the way down, Bullet looked at me and said, "At least Mike's big ass isn't bare, too." Once he was down and we found that he had no medical problems, we packed up and headed for the station.

After going through shift change, I headed home and crawled into bed with Teri and the dogs. It didn't take me long to go to sleep. I didn't even wake her up when I got in. And I slept so soundly that she didn't wake me when she got up.

When I got up, I went to the living room and told Teri about delivering the baby, the tac team, and the drug addict in the tree. She laughed about the calls but she had the biggest laugh when I told her what Brian Walker did to Chief Anderson. She's met Chief Anderson and she doesn't like him any more than we do. She said, "Couldn't happen to a nicer guy."

The rest of the weekend flew by. Unfortunately, it didn't last long enough. Before I knew it, it was time to get my things ready to go to work again. I had a good weekend spent with my favorite people in the world, but now it was time to go back to work. As I cuddled up to Teri, I decided not to dwell on it too much. I decided to just enjoy my time with her.

Chapter Twenty-Two

Today started just like every other has for the past week. Checking hydrants. We worked all day on them and knocked out a total of seventy-five. Not a bad total for the day but not so many that we wore ourselves out.

That evening, we watched another movie in the classroom. Mike Berrier absolutely refused to watch a horror movie tonight. I can't understand why. I thought added to his enjoyment of the movie the other night, but I guess I was wrong.

While we were watching the movie, when anyone would walk past Mike, they would grab his shoulder and yell, "Boo!" He didn't take it too well. After the fourth time, he finally headed to the dorm. He decided he had had enough teasing and decided to go where he would be left alone.

When he got to the bedroom, he found that his bed had been rigged. He found out the hard way. We had set his it to fall apart when he laid down. It worked like a charm. He got up, cussing everyone in the station and put the frame of the bed back together. He lay down again and when he went to pull the covers up, he found out that his bed had been short-sheeted. He fixed that and lay down again. This time, he found that someone had filled his pillow full of talcum powder. He got up and was mad as hell. He went stomping into the laundry room and threw the pillow and case into the washing machine and turned it on.

Fortunately for Mike, he had an extra pillow in his locker. Unfortunately for Mike, after putting the pillow on his bed, he left the room. This time we used a water balloon. He cautiously lay down and slowly pulled the blankets up. Once he was convinced that there was nothing wrong with the bed, he laid his head down and the balloon popped. His head was soaked and he was cussing us for all he was worth. He was getting ready to start rigging beds around the station when the ladder truck was called out to a house fire. By the time they returned to quarters, everyone was in bed, which just makes it impossible to set them up.

While the ladder truck was out, Engine 18 and Squad 18 were dispatched on a seizure call. When we arrived, we found a huge man lying on the floor in the hallway. As you enter the house, there is a hallway to the left going from the front door to the kitchen. Or you can go straight in to the living room and there is another doorway on the left end of the living room to the kitchen.

We checked vitals on the patient and he was breathing normally and had a good, strong pulse. EMS arrived and took over care of the patient. The hallway is narrow so they took the stretcher into the kitchen. The two paramedics struggled to carry the man into the kitchen but managed to get him on the stretcher. Then he woke up and panicked. He threw a six-inch punch to the sternum of one of the paramedics and damned near knocked him out. They both backed off quickly. I was standing in the kitchen and ended up getting blocked in. Bullet and Rick took off out a side door of the kitchen. They slid the kitchen table in front of me so they could get out. I couldn't get to the door.

On the other side of me was the wife, who wasn't about to leave. It turns out that the patient was a former college football player and as big as a house. His wife, on the other hand, was barely five feet tall and may have weighed ninety pounds soaking wet. Once he saw me and started coming towards me, I picked his wife up and carried her into the living room. It was either that or run over her.

When I got to the living room, he came down the hall and blocked my exit through the front door. I made sure I kept the coffee table between the two of us and I was holding the steel oxygen bottle. He came at me and threw a punch at my head. I was able to block the punch with the oxygen bottle, but only by pure luck. He stood there staring at me while shaking his hand. I said, "Hurts, doesn't it?"

He said, "Yeah!"

"Well, now we have an understanding. You hit me; I'm going to hit you with this bottle. It already hurt your hand. Just think how it'll feel on your head."

He turned his attention back to the paramedics and chased them into the kitchen. That was my cue to get the hell out of Dodge. I bolted out the front door and barely escaped his reach. I hated to leave the paramedics behind, but I didn't see any reason to get my ass kicked, too. I went down to the street and met Bullet and Rick down there. They asked me if I was okay and I said, "Yeah, no thanks to you two. You could have left me room to get out of there."

We stood around waiting for the paramedics to get out. As I said, I hated to leave them in there, but on a call my safety is first, my partners' safety is second and the patient comes in third. I know that may sound selfish to some people, but if I get hurt, I am no good to anyone else on the call. Besides, it makes your career last a little longer.

The paramedics finally realized that there was nothing that they could do. The patient wasn't even listening to his wife. He was the head of a Christian athletic organization in town but he didn't care about that either. All his wife could say was, "Honey, let's pray." At that point, he didn't even know that she was his wife, much less what he did for a living.

The police arrived and they had to wait to take him into custody. He wasn't a criminal, he was a patient. They couldn't just go in and take him. He was standing on the front porch, directly in front of a full glass storm door. If they tried to take him there, he could have been hurt by the door. After half an hour, one of the police officers finally lured him away from the front door by pissing him off. Once away from the door, the patient was pepper sprayed, taken to the ground and cuffed. Apparently the pepper spray made him snap out of his rage and he calmed down instantly.

We washed his eyes out with a garden hose next to the house and he was loaded into the ambulance. Later we found out the reason for his seizure and subsequent erratic behavior. He had a brain tumor the size of an orange. Up until his seizure, he never showed any signs and never had a problem controlling his temper. He was operated on within a week and made a full recovery.

We made it back to quarters and I called Teri and gave her my love. It was time to crash for the night and I hoped that we didn't have any calls. An hour after turning in, I realized that it would not be a quiet night.

We were dispatched to a fire at the city jail. When we arrived, the prisoners were rounded up and taken to another location in the jail. One of them was mad and set fire to his bed. He was probably pissed off because the salad fork wasn't chilled at dinner. Who knows. These guys may be locked up, but they still have a pretty good life. They get three free meals a day. They play basketball or lift weights. They have better cable than I have at my house. They are living on top of the world. They may have to worry about Big Bubba turning them into his bitch, but they live better than we do. Not to mention the fact that they have more rights than we do.

We were able to knock the fire out easily. We just took a hose from one of the hose stations in the jail. We didn't have to dirty our equipment. Then the head of the jail got the bright idea that we should go from cell to cell to see if there were any other fire hazards that needed to be eliminated. I said, "Yeah, don't let them have matches or lighters."

It didn't do any good. We spent the next two hours going through each cell and looking for ways to start fires. I was pissed. We had done our job. We were there to put the fire out. It was not our responsibility to help do a shake down. We found matches and lighters and the guards wouldn't let us take them. They had to have a way to light the cigarettes that the taxpayers were buying for them. After not letting us do anything, I told the crew to skip it and go back to the truck. We were out of there. The head guard was pissed and threatened to call my boss. I told him, "Go ahead and call him. It's your job to do shakedowns, not ours."

He said, "What the hell is his number?"

I looked him straight in the eye and, without cracking a smile, said, "9-1-1."

We got back to the station and I did the report and headed back to bed. We were able to sleep the rest of the night, but just barely. Right after we got up at 7:00, we were toned out to the scene of a rape victim. I don't like these calls at all. It's not that I have anything against a rape victim; it's the fact that most women who are raped cannot trust men. Our entire crew is made up of men. The only thing you can do in this situation is try.

When we entered, there were two male police officers and they were keeping their distance. Every time they tried to go near her, she went nuts. I tried to help her but she wouldn't let me touch her. I finally got an idea. I laid some four-inch by four-inch gauze pads on the kitchen counter along with gloves and a roll of gauze. She had a deep cut on her forehead and it needed to be covered. I stepped back and asked her if she could put one of the four-by-fours on the cut on her forehead. She said she could do that so I asked her to put the gloves on first. She did and placed it over the cut. I then had her take her own pulse and tell me the results. I took off my watch and slid it across the counter so that she could keep track of the seconds.

After writing down the results, I asked her to move her hand. The blood had soaked through the four-by-four so I had her place another one on top of the first. EMS arrived and there was a female on the truck. The female paramedic was the only one allowed to touch her. She gave me my watch back and I went outside to let the victim have some privacy. A female officer arrived and helped the paramedic walk the woman to the ambulance.

We went back in and cleaned our trash up. We rode back to the station and did our shift change. On the way home, I just prayed to God that this would never happen to Teri or Nikki. It would kill me to see them like that. I just prayed that God would watch over them and never let something like that happen.

I went to bed when I got home. I had most of my regular sleeping partners with me. Teri was at work, but it didn't take much convincing to get the dogs to get on the bed and go to sleep. They like to play, but they love to sleep. That's the only way to get them to calm down when I get home. I have to let them sleep on the bed with me. And it gives me some company.

After getting out of bed, I did some work around the house. I went out and did some yard work and fixed a leaking showerhead. After picking up Joey and Nikki at school, we worked on the yard some more. They don't mind working in the yard as long as they get paid for it. I was raking leaves and putting the leaves in the wheelbarrow and Joey would run them around to the back yard and put them in a pile. He had a pencil and a piece of paper and was keeping track of each load. We made an agreement that he would get a quarter per load of leaves. The only thing is, he didn't think about the distance from the front yard to the back yard. After two trips to the back yard, he said he needed a raise. I agreed.

I had to hand it to him. He has a great head for business. I just hope that when Teri and I are to old to take care of ourselves, he doesn't try to sell us to an adult diaper company as test models. Or sell us to science. If he could find a way to make money off of us, he would probably do it.

We worked in the yard for a few hours and the kids made a small fortune. We went in to start on dinner. I did steaks on the grill and we had a good dinner. When Teri came home, we feasted on steaks, potatoes, salad and corn on the cob. It was a good dinner and I ended up eating too much. But it was worth it. It was good, even if I do say so myself.

Teri and I turned in after the kids were in bed. She was tired from a hard day at work and I ended up helping her out with a full massage. She loves my massages. I just wish I could do that for her more often. It takes awhile and she will usually fall asleep when I give her one. But if I can make her feel better, it's worth all of the effort in the world. That's why I'm here. To make my wife feel better.

The next day was pretty uneventful. I finally took Danny Steele's cane to him. He almost cried when he saw it. He said he never expected to get something like that. He was very appreciative. After giving him the cane, we talked about retirement and how he was doing. It turns out that he is finished with the operations. A total of nine operations on the same leg. That's got to be rough. I asked, "How are you handling not being on the truck?"

He said, "I can handle not having to put up with the bullshit from city hall and the administration. I don't care if I never have to deal with any of those assholes again. The thing I miss the most are the guys."

"Well, why don't you come by the station once in awhile?"

"I just feel like I'm no longer welcome there."

"What? Why the hell would you feel like that?"

"It's just that you guys are still doing the job. You are still working and I can't do shit anymore."

"Danny, that's bullshit. There isn't a man out there that doesn't think you were screwed. And you're welcome at any station at any time. Just stop by."

"I know. It's just that I feel like I can't do my part any longer."

"You may not play a part in the fire department, but you're still a friend and that means more than any job."

"I don't know. I guess it's not the fact that I can't do the job. I just feel like some of the guys may feel like I took the easy way out."

"Don't even think like that. Any man that goes through all of the operations that you have been through definitely didn't take the easy way out."

"Maybe not. I guess it'll just take a little time before I can go back."

"Well, don't make it a long time, okay?"

I couldn't believe he would feel that way. I know that no one on my shift has ever felt that Danny took the easy way out. There hasn't ever been a bad thing said about him. The only thing that anyone has ever said about the whole thing is that the fire department should have done more for him after he was injured. They paid his hospital bills and they have paid for every operation. Well, more to the point, workmen's compensation paid for it. But the department could have put him in another position.

With the knowledge that he has, he would have made one hell of a training officer. At the time he was injured, we had several openings in the inspections department, but they didn't want to offer him any of those jobs. He was already certified as a level three inspector and all they would have had to do is transfer him. But the Fire Chief, Al Peters, didn't want to do that. Danny and the chief came on together and they never got along throughout their careers. I guess this way, Chief Peters had the last laugh. Of course, maybe that's the reason no one on this department likes him.

The kids and I went to the library and they worked on their homework while we waited for 6:00. We were supposed to meet Teri in the parking lot at the hospital and go out to dinner. They finished their homework early and we went outside and walked around until it was time to go. Then we drove over to the hospital and met her.

After dinner, we headed home and the kids went their separate ways until it was time for bed. Teri and I sat on the deck and talked about Danny and the way he felt about visiting the stations. She said it may take some time before he could go back with his head held high. I guess it has to do with pride. I think the best thing that could happen is for Danny to come by and visit once in awhile. Chief Peters may have had the last laugh as far as Danny's career goes, but Danny has more respect in retirement than Chief Peters will ever have as chief. Now all we have to do is convince Danny of that.

I fell asleep with the hopes that I would never feel like Danny. I also thought about the others that had left before him. Whether they left due to disability or after they had put their time in, they always come back to visit. I hope that one day Danny will come back.

Chapter Twenty-Three

We didn't have to do hydrants this morning. Instead, we had to go to our continuing education class for EMT. It's not the most exciting thing to do, but it kills a morning. Besides, it's starting to get cold outside and today was supposed to be in the low forties all day. There is nothing worse than working with wet hands on a cold day.

Before we left for our class, Bullet came in with a big bag of yellow squash. His father has a farm and sent a ton of it to him. Bullet decided to bring some in and cook i up for dinner. The only problem is that he had a fifty-gallon garbage bag full it. There was no way he could eat everything his father sent him so he brought it in and had someone off the other shift take some to the other stations.

While Bullet was washing the squash he kept for our station, Frank Marconi came in the kitchen and started looking at the squash. He is from New York and had no idea what Bullet was washing. He was inspecting a squash and asked, "What is this?"

Bullet said, "It's a squash."

"What do you do with it?"

Bullet stopped what he was doing and stared at Frank for a moment. Then he said, "What do you think you do with it? You eat it."

Tony kept studying the squash and asked, "Well, how does is grow?"

Without missing a beat, Bullet said, "On trees."

Of course, we all know that squash grows on a vine, but obviously Frank didn't. Frank fell for it hook, line and sinker. Bullet knew he had him. When Tony was out of earshot, Bullet got on the phone and called Station 15, where we were to have our EMT class that day. He talked to Mike Phillips and told him the squash story. The squash had already been delivered to Station 15 and they went outside and tied it to the trees around the station. Bullet then called a friend of his that lived up the street from our station. After we left for class, Bullet's friend came to the station and tied some squash to the trees around our station.

When we arrived, Frank saw the squash hanging from the trees but didn't notice the string holding them up there. He went inside and asked where they got their squash trees. They were very convincing. They told him that a year ago they dug up the old trees in front of the station and planted squash trees in their place. They told Frank that there was a nursery on the other side of town that sold them.

After our class, we went back to quarters and that's when Frank saw the squash hanging in the trees around our station. He started getting suspicious. He went outside and pulled one down. When he saw the string hanging from it, he went in and told Bullet, "I thought you said these things grew on trees."

Bullet said, "They do."

"Then why is there string tied to the end of this one?"

I looked at the squash and said, "I can tell you've never been to a farm. You have to tie the squash to the tree or the weight will pull it loose before it gets ripe enough to eat."

Frank thought for a moment, then said, "Oh. Okay."

That's when Bullet lost it. He started laughing and Frank knew he was had. I looked at Bullet and said, "You can take the boy out of the city, but you can't take the stupid out of the boy."

Frank threw the squash at Bullet and stormed off. We sat there laughing. Frank didn't like to be teased about being a city boy or a Yankee. When I first came out of training, if I remember nothing else that my first captain taught me, I learned that you never pass up a free shot. And it helped that Frank left the door wide open.

That afternoon, we braved the cool weather and went out and did some hydrants. It was miserable out there. The wind was blowing making everything worse. Rick and I turned about twenty hydrants and finally said to hell with it. We went back to the station and warmed up.

Bullet fixed several different squash dishes for dinner along with a pot roast. Frank came in the kitchen and got pissed when he saw all of the dishes. As he was turning to leave the kitchen, I took him by the arm and said, "Hey, don't take it personally. He had to fix it all or it would go bad."

We had dinner and went to the day room to watch television. We were watching Jeopardy and shouting the answers out. We would argue about the answers the correct one was given. It was fun while it lasted, but we were dispatched to a reported house fire in a nearby neighborhood.

When we arrived on the scene, there was no one in sight. It turns out that a passing motorist with a cell phone called in the reported fire. I hate cell phones. They have caused us more trouble since everyone started getting them than false alarms. If these people would just check things out before calling they would realize that their "good deeds" were nothing more than a mistake. This time, the house fire that the good citizen called in was actually steam coming from a drier vent in the back of the house.

We've had people call in with car fires that turned out to be overheated vehicles. Vehicle accidents with injuries that turned out to be small bump ups. Mulch fires that turn out to be nothing more than steam coming off the pile. In cool or cold weather, fresh mulch piles will generate heat from the decomposition of the mulch. It is warm inside the piles and it generates steam. I don't mind people calling in emergencies, I just wish they would check things out beforehand.

And it doesn't only happen to the fire department. The police get their share of false calls. I remember one call they ran that turned out to be nothing. Someone called in a burglary in progress in one of the higher priced neighborhoods in town. Several police officers arrived on the scene at the same time with their guns drawn, coming in like storm troopers. It turned out that the men "robbing" the house were actually movers and the owners nearly went into cardiac arrest when the police arrived.

We returned to quarters in time to catch the beginning of the college football game. It was supposed to be a great match-up but I didn't give a shit about either team so I decided to go play on the Internet for a while. I go on-line to look up carving patterns. You can find all kinds of patterns on the internet if you know where to look for them.

After spending some time on the computer, I called Teri and gave her my love. I told her what we did to Frank Marconi and the squash. She loved it. She did say it was cruel but she loved it. I said goodnight and went to bed.

At 3:30, the tones went off and we were dispatched to the scene of a vehicle accident with injuries. En route, fire communications informed us that the accident involved a tractor-trailer that had overturned on a vehicle. When we arrived on the scene, we found that the tractor-trailer didn't turn over, but the backhoe that was on the trailer had broke free of the chains and turned over on the minivan that was in the next lane.

We surveyed the scene and found fluids leaking all around the minivan. The fluids were leaking from both the van and the backhoe. We dammed up the drains surrounding the area and built another dam all the way across the road, down hill from the wreck. We then started working on getting the backhoe off the van.

We were able to get two airbags under the backhoe on either side of the van. We then inflated the airbags as a tow truck on the other side of the trailer pulled it upright and away from the van. Once the backhoe was off the van, we were able to look into it for the first time. When I looked inside, I saw the driver was badly injured and the passenger sitting next to him was nearly cut in half at the waist. We then moved on to the back seats and found two children. Amazingly, the children were unharmed by the accident.

We were able to pop the rear doors open and get the two children out. Once they were away from the scene, we set about getting the two victims in the front out. I was able to check for a pulse on both of them. The female passenger was dead. There was nothing we could do for her. We could have extricated her and tried to revive her, but when I felt a pulse on the driver, we decided to get him out first. He had a good strong pulse and, even though it appeared that he had severe head trauma, his blood pressure was good.

We were able to get the spreaders inside the door and pop it open. We moved him outside and EMS started working on him. One of the paramedics on the scene checked the female again and said there was nothing that could be done for her. When a person dies, the blood in their body pools to the lowest lying parts. When he checked on her, blood was pooling in her feet and he knew that she was dead. There was no heat coming from her body. Had she been warm, we would have tried to get her out quick. There's a saying in EMS, they're not dead until they are cold and dead. If there's no pulse, but the victim is warm, or generating body heat, there is a chance to save them. In this woman's case, not only was she cold and dead, blood was pooling and rigor mortis was setting in.

We packaged the man for transport and he was off to the hospital. While we were getting him out of the van, the children were taken by another ambulance to be checked out. Once the man was out of the way, the coroner arrived and we extricated the woman. It took awhile to get her out and we had to wear gowns because of all of the blood. Once she was gone, we still had to wait for the roll back to come get the van. Once it was loaded, we had to drain the fuel out of the tank and make sure it would not be leaking all the way to the junkyard.

We waited around until the environmental cleanup agency came to the scene and started cleaning up the mess. Not only did we have gas on the ground from the van, there was also diesel fuel and hydraulic fluid from the backhoe. While they were cleaning that up, we started looking at the backhoe. A state trooper on the scene showed us where the chain broke and allowed the backhoe to fall over when the truck was going around a curve. There was no way for the driver of the truck to know that the chain was broken. He was so upset, he was about to pass out. Nothing like this had ever happened to him before and he felt like everything was his fault. But according to the state police, there was nothing he could do. The driver ended up going to the hospital with chest pains.

I felt sorry for the guy. He was just doing his job, moving the backhoe from one building site to another. The problem is that, due to faulty chains, a fatal accident occurred. It was not fair to the victims of the wreck. It was not fair to the driver of the truck. It was not fair to anyone involved in working at the wreck. But shit happens. I realize this may sound very callous to people, but you have to look at it that way. If you sit around dwelling on all of the bad calls you run and think about all of the people that you cannot help, you'll burn out in this job very quickly. Maybe that's why emergency services employees have such a warped sense of humor. If we didn't, we would go nuts.

We finally got back to the station in time to get a cup of coffee before shift change. I went home to go back to bed. Just me and the dogs. We had a good nap, then went outside to get some exercise. I threw the ball around for them and they would bring it back to me. The problem is, they have a tendency to be ball hogs. They never want to give it back. But I have their number now. I take two balls.

They tired of playing fetch so I went to the shop and started working on a new cane. I didn't have anyone in mind to make it for but I figured I would make one for the hell of it. I like to carve and I like to make canes. I'm not sure what it is, but they are fun to make. And I have found that people like to use them when I do give them away. They aren't heavy but they are strong enough to hold up under the weight of even the largest person.

I went and picked the kids up and brought them home. They went outside to play and I went to my studio to play. I like getting lost in my music and it gives me a lot of pleasure. It relaxes me but I do have a tendency to lose track of time. I didn't realize that Teri had come home and already started cooking dinner. I didn't realize it until I smelled the food.

I went in the kitchen and gave her a hello kiss. She didn't mind that I hadn't started dinner. She said it was okay because I was playing for my dinner. She likes to sneak in on me when I play. She says it's different when I don't know that she's around. She says that it's like I enjoy it more when I play for myself than try to play for someone else. She says both are good, but when I play what I want to play, what I call "fiddle farting around," it's different somehow. I've never noticed a difference. I was just glad that she wasn't pissed that I hadn't cooked dinner.

We had dinner and went into couch potato mode in front of the television. Not a bad way to spend the evening. Once we decided there was nothing on television, we put a movie on. After putting the kids to bed, Teri and I decided to play some music of our own. That's one good thing about our kids. They could sleep through a bomb going off. They don't wake up and disturb us when we are "having fun."

We fell asleep in each other's arms, but at an angle. Jake was persistent, trying to get in between us. We just kind of angled our bodies so that we could stay in contact but he could be where he wanted to be. I slept well that night and the next morning got out of bed before Teri got up.

I went to the kitchen and cooked breakfast. I cooked what I call the firehouse breakfast. Bacon, sausage, eggs, biscuits, grits, gravy and antacids for desert. They like it. I also like to call it the Heart Blocker Special. It goes over big with Teri and the kids. And the best part is that when I make breakfast, they volunteer to clean up the mess. That way I get to take the rest of the morning off.

We went out and did some shopping for the kids. I found a couple of things I needed for the shop. Teri found some things that she needed for the house, but most of the things we bought were for the kids. They needed some clothes. They're both growing like weeds and getting bigger all of the time. I can't believe it. It doesn't seem like it's been that long since they were in diapers but I guess it has. The next thing I know, we'll be buying Nikki a wedding dress. But as long as she doesn't marry some shit head, I guess things will workout all right.

We came home and did about the same thing we did last night. Sat around watching movies. After the movies, I got my stuff ready for work the next day. I hate working

Sundays, but I figured I would get one out of the way and not have to worry about it for another three weeks. I fell asleep with Teri and the dogs and had a good night's sleep.

Chapter Twenty-Four

I went in early today. I got to work at 7:15 instead of 7:45 like I normally do. I was able to get a couple of cups of coffee before we did shift change. We sat around talking about a few memos that had come out the other day.

At 8:00, we did shift change and checked the truck out. You have to do that each day you come in to make sure that you have everything. You never know if one of the other shifts might have left something behind on a call so you have to check to be sure that everything is there. Bullet, Rick, Tony and I checked over the truck to make sure we had everything and that the oxygen bottle was full and the water tank was topped off. Once we did that, we headed to the kitchen to wait on breakfast.

We had barely finished breakfast when the tones went off. We were dispatched to a reported building fire at a nearby office complex. The offices were built in a remodeled factory built in the late 1800s. It was a nice building and a landmark in the city. It was a shame when we pulled out of the station and I had to report that we could see heavy smoke in the direction of the building all the way from our front ramp.

When we arrived on the scene, a man standing on the sidewalk across the street from the building informed us that there was a woman who was inside the building. His wife went in to work on Sunday morning to catch up on paperwork and he knew that she was still in the building. I reported this to fire communications and we got ready to go in. I didn't like the looks of it, though. Fire was coming out of every window on the top two floors. This fire must have took off pretty quick, for no one took notice it until now.

When Engine 15 arrived on the scene, they prepared to be our rapid intervention team. That means that the entire time we are in the building, they have men in air-packs and on a hose line to come in and get us if anything goes wrong. For every man we have inside, there is one man outside. Before we entered the building, I checked with the husband to get directions to her office.

We entered the building and headed up the stairwell. We were told that her office was on the second floor, directly over the front door of the building. We made our way to the top of the stairs and found a lot of heat radiating through the door. By the time we had worked our way up there, Chief Williams had another crew, Engine 26, join us inside. We were able to get the door open and immediately hit the ceiling of the hallway. That hardly made a dent in the fire. We were about to give up, figuring the lady was dead when Chief Williams came on the radio and said that a woman had thrown her chair through a window on the second floor above the doors of the building. He had the ladder crew bring a ladder to get her down.

I knew that it would take some time to get the ladder raised and get her down from the second floor. The way this building was going up, it might be too late when they finally do get up there. We opened the door enough to get the hose line through and started working it on the ceiling in the hallway. After about twenty seconds, we were able to knock the fire back enough to get into the hallway and start hitting the fire with a more effective stream. We were taking a lot of heat at the moment but we were able to be effective at least a little in holding the fire off from the office where the woman was.

I got on the radio and asked Chief Williams how soon it would be before the woman was out of her office. He said that the ladder crew was heading up to get her at that moment. I told him to have them step it up a little. We were taking a lot of heat and needed to get the hell out of there. We were starting to approach flashover levels. A flashover means everything in the room reaches ignition temperature at the same time and it comes at you like a firestorm. I didn't want to be around when that happened.

After another thirty seconds, Chief Williams informed us that the woman was on her way down the ladder and ordered us to back out. We were working at two different angles. Rick, Tony and I had our hose stream pointed one way down the hall to keep the fire off of all of us. The crew from Engine 26 was concentrating the hose stream on the area in front of the woman's office. As soon as we were told to back out, I had Engine 26 go first. Most of the heat and fire was coming from the direction of our hose stream and I wasn't about to let that go. They backed out and headed down the steps as we started to back out. I had Tony go to the next level and start pulling slack on the hose line.

Rick and I managed to get the nozzle back in the stairway and started heading down the steps to exit the building when everything on the second floor took off at once. The force of all of that fire in the hallway blew the doors open in the stairwell. It turned the stairwell into a blast furnace in a matter of seconds. Rick was ahead of me by about eight steps. When the heat hit me in the back I dove to the bottom of the steps. Actually, I'm not sure if it was a dive, the fire knocked me down, or if it was a combination of both. I just know that in a split second I went from running down the stairs to bouncing off the wall at the bottom and rolling down the next set of steps.

Fortunately, it was much cooler down there than the next level of steps. The steps were shielding us over our heads. I tried to get up. I couldn't move my left leg. No matter how hard I tried, I could not stand up. Rick came back and found me. He helped me up but I couldn't put any weight on my left leg. He finally told me to get on his back and ride piggyback out of the building. With all of the bouncing, it hurt like hell. But I was out and I was alive.

When I got outside, the men around the building swarmed all around me. Bullet got someone to take over on the pump for him and he came over to see if I was okay. All of the pain in my leg was in the thigh area. They took off my helmet, air-pack and coat. Then they began cutting my pants off. When I looked at my helmet, I couldn't believe it. The back of the helmet was blistered and there was a chunk of wood stuck in it. Two nails held the wood there. This is a good example of why we need the helmets.

Once my pants were cut off, I noticed a large deformity in my thigh and I was told that my left leg was about three inches shorter than my right leg. Femur fracture. This was going to suck. EMS arrived and Jack Walters and James Stokes put the hair traction splint on me. I've had the splint on before, but only for EMT class. Every time you put it on one of your coworkers for practice, you have a tendency to tighten the straps too tight and make it as uncomfortable as possible for them. That's just part of the game. But when Jack and James put it on me and pulled the broken bones of my femur back into alignment, the relief I felt was damned near as good as sex. Not quite, but pretty damned close.

Once they had splinted my leg, I was loaded into the back of the ambulance and taken to the hospital. I was just glad that Teri wasn't working today. She always told me that my being injured and coming to her emergency room was one of her biggest fears. I was going to the same emergency room, but she was off. When we got in the back of the ambulance, I asked Jack, "Well, how does it look?"

He said, "Same as usual. You still have the ugliest legs I've ever seen."

"Well, as long as nothing's changed."

"Dalton, I'm not going to lie to you. It's a pretty bad break. But don't worry about it. We'll take good care of you."

"I know you will. Because I know where you live."

Once I arrived at the hospital, I told the nurses in the ER that they had better not call Teri. I wanted her to hear it from me, not one of them. I knew that if she heard it from me, it would make her feel a little better than if they called her. When they finished examining me and had taken x-rays, I was moved close to a telephone and I called my father first. I had him go to the house to get Teri. I didn't want her to try to drive in. She worries about me so much and I didn't want her to try to drive. Pop headed to the house while I called Teri.

I said, "Hey. I need you to stay calm."

She asked, "What happened?" She sounded worried the minute I said to stay calm.

"I fell at work and broke my leg."

"How bad is it?"

"Well, the doctor says I'll have to have surgery on it. It's a femur fracture, and he said that I'll be going under the knife in a little while."

"I'm on my way."

"Hey, don't try to drive up here yourself. Pop's on his way and he's going to drive you up here. Promise me you won't try to drive."

"I promise. But he had better hurry."

"He's on his way now."

By the time I told her what happened and how I fell, Mom and Pop were there waiting on her. They brought her to the ER and she came back to see me. Joey and Nikki were in the waiting room and wanted to see me, but they aren't old enough to come back to the ER. I finally talked the nurses into putting me in a wheel chair so I could go out and see them. Once I was able to joke around with them a little bit, they felt better. I'm not sure how Teri felt at that point, but all I could do is give her my love.

After another hour of waiting around, I went under the knife and they did surgery on my leg for over three hours. I woke up in the recovery room, sweating like a pig and feeling like shit. Teri was there waiting for me to wake up. I hated that she had to worry about me so much. I never wanted it to come to this. I love my job, but I love her with all of my heart. Teri is my reason for living. She is my heart. I hated to make her worry about me so much.

When I finally was able to focus enough to know who was sitting beside my bed, I said, "Hey, baby."

She kissed me on the forehead and asked, "How do you feel?"

"I'm not in pain or anything. I just feel like shit. Drained. Hell, I'm having trouble moving my head right now."

"Well, don't worry. That's normal. You'll feel better after you wake up a little more."

She was right. When I finally had all of the anesthetic out of my system, I felt a hell of a lot better. By that time I was moved to another room where I would stay overnight. When the nurses finally left me alone, I called the station.

Bullet answered and wanted to know what was going on. "We've tried calling the hospital and they wouldn't tell us anything. They wouldn't let us see you when we stopped by this afternoon."

When he said, this afternoon, I realized it was 7:30 in the evening. I didn't realize it had taken this long. The fire call came in at 9:30 that morning. I told Bullet about the surgery and that I hadn't had a chance to talk to the doctor so I didn't know how long it would be before I would be able to come back to work. I said, "I guess I'll be laid up for awhile and hopefully go home in a day or two."

He told me to take care and said that he would let everybody know what was going on. He said that he would come by and see me tomorrow when he got off work. He also told me to pass along a message to the nurses. "You tell them that if they don't let me in to see you tomorrow, I'm gonna pop a cap in their asses." I said that I would pass the message along.

Mom and Pop walked in with pizza. Teri and the kids were there too and the six of us ate pizza and talked for a while. The head nurse walked in to check on me and was shocked that we were eating pizza. It turns out that there was a shortage of beds in the area of the hospital I was supposed to go to. It took her by surprise because they gave me a bed in the cardiac ward. She had just come on duty and no one told her that I was there. It took some convincing, but Teri finally explained the situation to her.

Finally, the surgeon came in to tell me about the operation. He said that everything went as good as could be expected. He told me that the damage was repaired and that I would be in the cast for about six months. Apparently it takes longer for the femur to heal than any other bone in the body. I asked him, "Well, how long until I can get back on the truck?"

He looked at me for a moment, then said, "Dalton, I don't know any other way to say this, so I'm just going to say it straight out. You won't be able to go back to the truck. The damage to your leg is just too much for you to go back to fire fighting. From here on out, you need to consider yourself a retired firefighter."

When I heard that, I felt like someone had hit me in the chest with a sledgehammer. I didn't know whether to shit or go blind. I had built my life around fire fighting. This was the job I was doing when I met Teri. It was the job I was doing when our kids were born. Being a firefighter isn't what I do, it's what I am. It's the only thing I have ever done, career-wise, that I have ever felt proud of. Now what was I going to do?

I still had my family, and I thank God for that every day. I am still able to carve and play music and I am also thankful for that. But I have never thought of not being a firefighter. I worked in television for a while after college and I didn't like it that much.

The day I became a firefighter, I never looked back. I know that I will never get rich being a firefighter, but I love the job. There isn't a career out there that I would want to do more than fire fighting. It's such a big part of my life that I can't even fathom not being able to do it anymore.

After the doctor left, I had Teri take the kids and go on home. I needed to be alone to think for a while. After they left, I started thinking about what I would do now. I had never thought beyond the fire department. I figured once I retired, that was it. I would hang around the house and play in the studio or do my carvings. I figured that I would retire because I wanted to retire, not because I was forced to. I never counted on disability. I had seen people go out on disability before, but it never occurred to me that I would have to go out that way myself.

I didn't sleep at all that night. I kept thinking about everything that had happened and if there was something I could have done differently that wouldn't have placed me in this position. I couldn't come up with anything. Every time I thought about not being on the truck, doing the job that I loved, it brought tears to my eyes. I was now able to realize what Danny Steele had been going through for the past several months.

Bullet dropped by to see me and it nearly floored him when I told him what the doctor said. He looked at me in astonishment and asked, "Well, what the hell are you going to do now?"

I said, "I don't know. You know there's nothing on the department for me now. After the way they treated Danny Steele, I'm out, too."

"Well, maybe the chief can come up with something."

"I doubt it. According to the doctor, I'll be in this damned cast for six months and then I'll have to go through rehabilitation. Shit, by the time I go through all of that, it'll be time for me to retire anyway. I guess I'll just take my disability and retire now."

"What are you going to do for a living?"

"Hell, I don't know. Maybe go back to television. Maybe just bum around the house until Teri retires and we'll get a camper and travel the country. Who knows?"

"Just don't let the city fuck you over. You make sure they pay for everything. The operations, the rehab, everything!"

"Oh, you can count on that. I'm not going to let them off the hook."

"It's not going to be the same without you around the station. Man! Now I have to break in a new captain and set him straight on how things work."

"Well, just remember, I'm special. You may not have a captain that was a great as I was." He just laughed and said goodbye.

I had several visitors from the fire department and a couple from the city. The ones from the city were there to get the disability papers filled out. Once they are done with you, they are definitely in a hurry to get you the fuck out so they can get someone else in your place. I guess it's kind of like getting rid of the dead wood. Once they are done with you, you don't matter. I guess it's a good thing I'm not a horse. Otherwise, they would shoot me.

I finally got out of the hospital and went home. I'm not sure what the future will hold for me. I don't know where I will go from here but at least I am with the people that I love. I'm not so disabled that I can't be with my family.

I'm here with my beautiful wife, Teri, my son, Joey, and my daughter, Nikki. Sitting here looking at them, looking at my house, looking at the three dogs, and thinking about all that I have, I realized that there is life after the fire department. There is something to fall back on. My family.

At least, now I will be able to spend more time with my family. I won't have to be away at night and worry about them being in the house all alone. I'll be able to be there for all of the kids' school activities. No more working on Christmas. No more working on their birthdays. It will be nice to be there for my family. It'll be great. As long as I have the support of my family and friends, I'll get through this. They will get me through it.

It could be a good thing. No more death. No more destruction. No more overnight calls. No more calls when it's blazing hot or freezing cold. No more missed meals or cold dinners. No more fires. No more wrecks. No more calls at night, yanking you away from a good dream and a warm bed.

God! I'm going to miss this job!

Firefighter's Prayer

When I am called to duty, God, wherever
flames may rage, give me strength to save some
life whatever be its age. Help me embrace a little
child before it's too late, or save an older person
from the horror of that fate. Enable me to
be alert, and hear the weakest shout and
quickly and effectively put the fire out. I want to
fill my calling and to give the best in me, to guard
my every neighbor and protect his property.
And if according to Your will, I have to lose
my life, please keep with your protecting hand
my family free from strife.

Amen

About the Author

David Harris grew up in North Carolina and California. He graduated from Appalachian State University with a Bachelor of Science Degree in Communications in 1987. After working in television for seven years in various Virginia television stations, he decided to follow his dream and become a firefighter. In April 1994, he joined the High Point Fire Department in High Point, North Carolina. He now resides in Midway, North Carolina, with his wife, Theresa, her two children, Aaron and Ashley, and their two dogs, Jake and Lakota.